Some knowledge of the internal organisation and microscopic structure of plants is fundamental to an understanding of their morphology, physiology and evolutionary relationships. *Anatomy of Flowering Plants* provides a concise introduction to this subject, with chapters on stems, roots, leaves, flowers, and seeds and fruits, each illustrated with light micrographs, scanning electron micrographs and line drawings. Established data and areas of currently active research are brought together in an interesting, readable and contemporary analysis of the fascinating subject of plant anatomy.

Anatomy of flowering plants

Anatomy of flowering plants

An introduction to structure and development

PAULA RUDALL

Jodrell Laboratory
Royal Botanic Gardens
Kew

Second edition

CAMBRIDGE
UNIVERSITY PRESS

Published by the Press Syndicate of the University of Cambridge
The Pitt Building, Trumpington Street, Cambridge CB2 1RP
40 West 20th Street, New York, NY 10011–4211, USA
10 Stamford Road, Oakleigh, Victoria 3166, Australia

First published by Edward Arnold 1987
Second edition published by Cambridge University Press 1992

Printed in Great Britain at the University Press, Cambridge

A catalogue record for this book is available from the British Library

Library of Congress cataloguing in publication data
Rudall, Paula.
    Anatomy of flowering plants: an introduction to structure and
development/Paula Rudall.–2nd ed.
        p. cm.
    Includes bibliographical references and index.
    ISBN 0 521 42154 3 (pbk.)
    1. Angiosperms–Anatomy. 2. Botany–Anatomy. I. Title.
QK641.R84   1992
582.13′044–dc20   92-13493
                                                                        CIP

ISBN 0 521 42154 3 paperback

UP

Contents

Preface

Plant anatomy is a fascinating subject, now unfortunately often neglected in university courses, and sometimes regarded by students as old-fashioned, probably because of the wealth of reference works from the last century, and the simplicity of many of the basic techniques, in an age of complex technology. However, this apparent simplicity can be misleading; the interpretation of prepared material requires some knowledge and experience. Often the contemporary plant anatomist has recourse to modern equipment such as the scanning and transmission electron microscopes, or refined methods such as the use of cinematography to analyse serial sections. Although a great deal of excellent work in plant anatomy was undertaken in the nineteenth and early twentieth centuries, particularly in Germany, this century has seen an enormous progression of research in many countries of the world, such as the United States, Brazil, Mexico, Russia, India, Germany, the Netherlands and the United Kingdom. At the present time much active research ensures that plant anatomy is still a lively field. This book sets out to present a basic introduction to the subject to students of botany and related disciplines. Different aspects of development are discussed and some emphasis is given to areas of currently active research, often with controversial interpretations, to try to capture the interest of the potential researcher. It is impossible to give a fully comprehensive account of such a broad subject in a book of this size, and examples of recommended further general reading are given on page 99.

I am very grateful to the Royal Botanic Gardens, Kew, especially the staff of the Anatomy Section, Jodrell Laboratory, for their continued support and encouragement.

1

General plant structure

1.1 Organisation

At maturity higher plants consist of several organs, which in their turn are made up of tissues. Simple tissues, such as parenchyma, are composed of groups of similar individual cells, and may be interspersed with other cell types (idioblasts). Complex tissues, such as xylem, consist of more than one cell type.

Three main vegetative organs are widely recognised: root, stem and leaf (Fig. 1.1). Roots are usually found in the soil, from which they extract moisture and nutrients, although there are many examples of plants with aerial roots. Most dicotyledons have a tap root with side branches (lateral roots); the tap root develops from the seedling radicle. However, in monocotyledons the seedling radicle commonly dies at an early stage; the roots of the mature plant are stem-borne (adventitious roots), and may be branched or unbranched. The stem and leaves, which together comprise the shoot, typically occur above ground level, although some stems are modified into underground perennating or storage organs such as corms or rhizomes, and underground bulbs have swollen leaves or leaf bases. Each leaf subtends an axillary bud, and the pattern of arrangement of the leaves on the stem is termed phyllotaxis. At the onset of flowering, the shoot apical meristem changes from a vegetative to a reproductive apex and subsequently produces flowers (chapter 5). Flowers are classically interpreted as modified shoots, and the floral parts (sepals, petals, stamens and carpels) as modified leaves borne on a central axis.

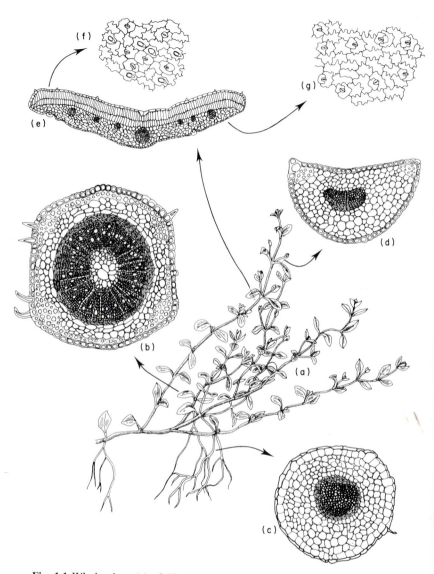

Fig. 1.1 Whole plant (a) of *Thymus* sp. (thyme) with cross sections of vegetative organs; (b) stem, (c) root, (d) petiole, (e) leaf, (f) adaxial leaf surface, (g) abaxial leaf surface.

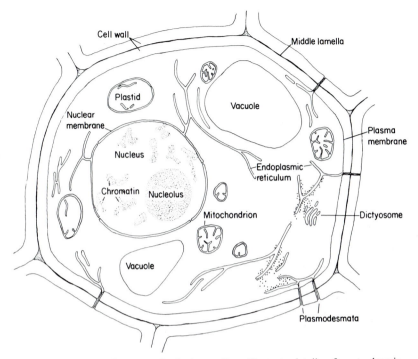

Fig. 1.2 Diagram of a generalised plant cell to illustrate details of protoplasmic contents.

1.2 Cell structure

Plant cells typically have a cell wall containing a living protoplast (Fig. 1.2). The layer between the cell walls of the neighbouring cells is termed the middle lamella. Many cells, when they have ceased growth, develop a secondary cell wall which is deposited on the inside surface of the primary wall. Both primary and secondary walls consist of cellulose microfibrils embedded in a matrix and oriented in different directions. Secondary walls are mostly cellulose, but primary walls commonly contain a high proportion of hemicellulose in the gel-like matrix, affording a greater degree of plasticity to the wall of the growing cell. The secondary wall may also contain deposits of lignin (e.g. in sclerenchyma cells) or suberin (e.g. in many periderm cells), and often appears lamellated. Thin areas of the primary wall, which usually correspond with thin areas of the walls of neighbouring cells, are primary pit fields, and usually have protoplasmic strands (plasmo-desmata) passing through them, and connecting the protoplasts of neighbouring cells. The connected living protoplasts are sometimes called the symplast. Primary pit fields often remain as thin areas of the wall even after a secondary wall has been deposited, and are then termed pits, or pit-pairs if there are two pits connecting adjacent cells. Pits may be simple, as in most parenchyma cells, or bordered, as in tracheary elements. In simple pits the pit cavity is of more or less uniform width, whereas in bordered pits the secondary wall arches over the pit cavity so that the opening to the cavity is narrow. Through a light microscope the outer rim of the primary pit field appears as a border around the pit opening (Fig. 1.8).

The cell protoplast is contained within a plasma membrane. It consists of cytoplasm enclosing bodies such as the nucleus, plastids and mitochondria, and also non-protoplasmic contents such as oil, starch or crystals. The nucleus, which is bounded by a nuclear membrane, often contains one or more recognisable bodies (nucleoli) together with the chromatin in the nuclear sap. During cell division the chromatin becomes organised into chromosomes. Most cells have a single nucleus, but examples of multinucleate cells include the non-articulated laticifers found in many plant families, such as Euphorbiaceae (Fig. 2.3). These laticifers are coenocytes, or groups of cells that have undergone division without corresponding wall formation.

Mitochondria and plastids are also surrounded by membranes. Plastids are larger than mitochondria, and are classified into different types depending on their specialised function. For example, chloroplasts are plastids that contain chlorophyll, for photosynthesis. They occur in all green cells, but are most abundant in the leaf mesophyll, the primary photosynthetic tissue (section 4.3). Membranes occur widely through-out the cytoplasm, sometimes bounding a series of cavities, as with the endoplasmic reticulum, or dictyosomes, which are associated with secretory activity. Vacuoles are cavities in the cytoplasm, usually colourless and containing a watery sap. They vary in size and shape in the different cell types, and often vary considerably during the life of a cell (see also Burgess, 1985).

Many cells have non-protoplasmic contents such as oils (which usually appear as droplets), mucilage, tannins (which are amorphous, and yellow, red or brown in colour), and starch granules and calcium oxalate crystals, both of which take many forms. Oil is produced in secretory idioblasts, often larger than adjacent parenchymatous cells.

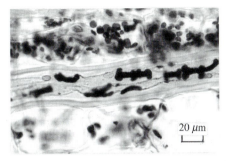

Fig. 1.3 *Monadenium ellenbeckii.* Elongated I-shaped starch grains in laticifer, with "normal" ovoid starch grains in adjacent parenchyma cells.

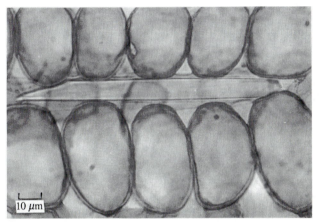

Fig. 1.4 *Crocus carpetanus.* Styloid crystal in leaf mesophyll.

Baas and Gregory (1985) and Gregory and Baas (1989) reviewed the literature on oil and mucilage (slime) cells in dicotyledons, and suggested that the two cell types are homologous. Starch is especially common in storage tissues, and the starch granules, which often appear layered due to the successive deposition of concentric rings, are formed in plastids (amyloplasts). Starch granules sometimes have characteristic shapes in different plants; for example, Mahlberg (1975) demonstrated that in *Euphorbia* and related genera, starch grains in laticifers are elongated compared to the more rounded starch grains of neighbouring parenchyma cells, and often of characteristic shapes (e.g. rod-shaped or bone-shaped) in different groups of species (Fig. 1.3). Calcium oxalate crystals may be prismatic or compound, and solitary or in groups. Styloids are long prismatic crystals (Fig. 1.4); raphides are thin, needle-like crystals, often in aggregates; druses (cluster crystals) are compound structures; and crystal sand is a group of many minute separate crystals in one cell (Franceschi and Horner, 1980). A few monocotyledon families, such as Gramineae, Cyperaceae or Palmae, sometimes have characteristic silica bodies contained in certain cells, such as idioblastic epidermal cells, and certain dicotyledon woods have silica bodies in ray cells.

1.3 Tissue distribution and origin

Complex tissues can be divided into three main groups: epidermis, ground tissue and vascular (conducting) tissue, each distributed throughout the plant body, and often continuous between the various organs.

The epidermis is the outermost layer of cells, covering the entire plant surface. It is a primary tissue derived from the outermost layers of the shoot and root apical meristems, although anticlinal divisions (at right angles to the surface) may occur in mature epidermal cells to accommodate stem or root thickening. In older stems and roots the epidermis often splits and peels off following an increase in thickness, and is replaced by a periderm (section 2.4; Fig. 2.14). In some monocotyledon roots the epidermis is worn away by the soil, and is replaced by an exodermis, formed by cell wall thickening in the outer cortical layers. The epidermis includes many specialised cell types, such as root hairs (section 3.2.1), stomata and trichomes on stems and leaves (section 4.2), and secretory tissues such as nectaries on various above-ground organs, particularly flowers (sections 4.2.4, 5.6).

Ground tissue often has a mechanical function, or it may be concerned with storage or photosynthesis. In general it consists of parenchyma, sclerenchyma or collenchyma (Fig. 1.5), often interspersed with idioblasts and secretory cells or canals. It forms the bulk of primary plant tissues and occupies the areas within the epidermis that are not vascular tissue or cavities. Ground tissue is initially formed at the apical meristems but may be supplemented by intercalary growth, and in monocotyledons by tissues differentiated from primary and secondary thickening meristems. In dicotyledons the ground tissue of secondary xylem (wood), formed by the vascular cambium, consists of fibres and axial parenchyma (section 2.3.1). The central area of ground tissue in older stems and thick leaves often breaks down to form a cavity (Fig. 2.2).

Vascular tissue, or conducting tissue, consists of xylem and phloem, and may be primary or secondary in origin. Primary vascular tissue is derived from procambium, itself produced by the apical meristems, and also by the primary thickening meristem in stems of monocotyledons (section 2.3.2). Secondary vascular tissue is derived from the vascular cambium in dicotyledons, and from the secondary thickening meristem in certain monocotyledons (Fig. 2.13), although secondary growth is rare in monocotyledons. Both xylem and phloem are complex tissues, composed of many different cell types. Xylem is primarily concerned with water transport and phloem with food transport. Distribution of vascular tissue varies considerably between different organs and taxa.

1.4 Meristems

Meristematic tissue consists of thin-walled, tightly packed living cells which undergo frequent divisions. Most of the plant body is differentiated at the meristems in well-defined zones, although cells in other regions may also occasionally divide. There are some remarkable examples of fully differentiated cells giving rise to entire plantlets, notably on leaves of Crassulaceae, such as *Kalanchoe* (Steeves and Sussex, 1989).

1.4.1 Apical meristems

Apical meristems are located at the shoot apex (Fig. 2.1), where primary stem, leaves and flowers differentiate, and at the root apex (Figs 3.1, 3.2), where primary root tissue is produced. Subsequent elongation of the shoot axis may occur by random cell divisions and growth throughout the youngest internodes. This region of diffuse cell division is termed an uninterrupted meristem, and is continuous with the apical meristem. However, in some plant stems, particularly in Gramineae (grasses), most cell divisions contributing to stem elongation occur in a limited area, usually at the base of the internode, which is then called an intercalary meristem (Fisher and French, 1976, 1978). Both intercalary and uninterrupted meristems represent growth in regions of already differentiated tissues.

1.4.2 Lateral meristems

Lateral meristems are located parallel to the long axis of a shoot or root, most commonly in the pericyclic region, at the junction between vascular tissue and cortex. Examples of lateral meristems are the primary and secondary thickening meristems of monocotyledons, which produce both ground tissue and vascular bundles (section 2.3.2; Figs 2.12, 2.13), and the vascular cambium of dicotyledons, which produces secondary xylem and phloem (section 2.3.1; Figs 2.6, 2.7, 2.8). Adventitious roots are often formed in the root pericycle; in these cases the pericycle could be termed a lateral meristem. The phellogen or cork cambium is also a lateral meristem in the stem or root cortex, where it forms a protective corky layer when the epidermis peels away (section 2.4; Fig. 2.14). However, a phellogen may also develop in the region of a wound, or at the point of leaf abscission.

1.4.3 Meristemoids

Meristemoids are individual cells or localised groups of cells that are responsible for the differentiation of distinct structures. They generally occur within tissue that is otherwise not rapidly dividing; for example, stomata are formed by meristemoids in the expanding epidermal tissue. In many cases they are formed by, or undergo, unequal divisions, which are an important element of the differentiation of certain tissues. An unequal cell division, either in a meristem or meristemoid, results in a larger, less active cell, and a smaller cell with dense cytoplasm. Examples of unequal divisions include that of the microspore into a larger vegetative cell and smaller generative cell (section 5.4.2), the formation of a root hair initial (section 3.2.1), a phloem division to form a larger sieve tube element and smaller companion cell (section 1.5), and the division of an epidermal cell into two cells of unequal sizes, the smaller of which is the meristemoid that will divide to form the guard cells of a stoma (section 4.2.2.).

1.5 Tissue and cell types

1.5.1 Simple tissues

Parenchyma cells (Fig. 1.5) are usually thin walled and often polyhedral or otherwise variously shaped, sometimes lobed. They are the least

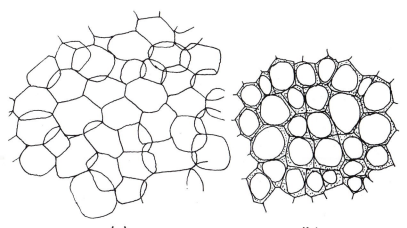

(a) **(b)**

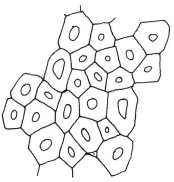

(c)

Fig. 1.5 Tissue types in cross section: (a) parenchyma, (b) collenchyma, (c) sclerenchyma.

specialised cells of the mature plant body as they most closely resemble meristematic cells. Indeed, in many cases they retain the ability to divide at maturity, so are instrumental in wound healing. Undifferentiated callus tissue is often produced at the site of a wound. A single isolated callus cell can be used to artificially grow a new plant using tissue culture methods. Parenchyma cells may occur in primary or secondary tissues. Relatively specialised types of parenchyma include certain secretory tissues and chlorenchyma, which contains chloroplasts for photosynthesis. Parenchymatous cells may be tightly packed or with many

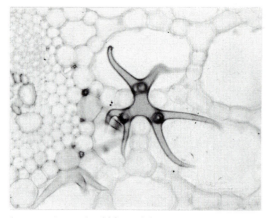

Fig. 1.6 *Nymphaea* sp. Astrosclereid in petiole aerenchyma. (x 250)

intercellular air spaces between them. Aerenchyma is a specialised parenchymatous tissue that often occurs in aquatic plants (hydrophytes), with a regular, well-developed system of large intercellular air spaces (Fig. 1.6). Cells with living contents that do not fit readily into other categories are often termed parenchyma cells.

Collenchyma (Fig. 1.5) consists of groups of axially elongated, tightly packed cells with unevenly thickened walls. This tissue has a strengthening function and often occurs in the angles of young stems (Fig. 1.1), or in the midribs of leaves, always in the primary ground tissue. Collenchyma cells differ from fibres in that they often retain their contents at maturity and do not generally have lignified walls, although they may later become lignified.

Sclerenchyma, also a supporting or protective tissue, consists of cells which have thickened, often lignified, walls, and usually lack contents at maturity (Fig. 1.5). Sclerenchyma cells may occur in primary or secondary tissue, either in groups or individually as idioblasts within other tissue types. They are defined as either fibres or sclereids, although there are transitional forms. Fibres are long narrow cells, elongated along the long axis of the organ concerned, and most commonly occurring in groups. Bast fibres are extraxylary fibres that occur in the cortex, and may be of economic use, e.g. in flax (*Linum*) or hemp (*Cannabis*). Sclereids may be variously shaped, and have been categorised into different types (Metcalfe and Chalk, 1979). Brachysclereids or stone cells are more or less isodiametric and may be present in any part of the plant. They include cells interspersed among paren-

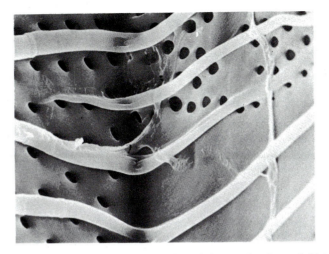

Fig. 1.7 *Tilia cordata* (lime). Inside surface of vessel element showing wall thickenings and intervascular pitting. (SEM, x 1800)

chymatous tissues that develop thick secondary walls as the plant ages. Macrosclereids are rod-like cells often found in the seeds of Leguminosae.

Astrosclereids are highly branched and sometimes star-shaped (Fig. 1.6), and osteosclereids are bone-shaped; both types are idioblasts found most commonly in leaves. During the development of the leaf they grow intrusively into surrounding intercellular air spaces or along middle lamellae, and their shapes are to some extent dictated by the nature of the surrounding tissues.

1.5.2 Complex tissues

Vascular tissue is complex and consists of several cell types: conducting cells, parenchyma cells, fibres and sometimes also sclereids, particularly in secondary phloem.

Xylem is the water-conducting tissue. The conducting cells of the xylem are called tracheary elements, and may form axially linked chains (vessels). They usually have thickened lignified walls and lack contents at maturity. Two basic types of tracheary element can be recognised: tracheids and vessel elements. Vessel elements have large perforations in their end walls adjoining other vessel elements, whereas tracheids do not. The perforations may have one opening (simple perforation plate)

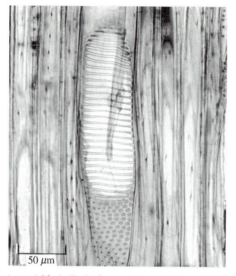

Fig. 1.8 *Alnus glutinosa* (alder). End of vessel element (in radial longitudinal section of secondary xylem), with scalariform perforation plate with numerous (*c.*29) bars, and bordered pits; adjacent fibres with simple pits.

or several openings, either divided by a series of parallel bars (scalariform perforation plate: Fig. 1.8) or by a reticulate mesh (reticulate perforation plate). Both tracheids and vessel elements may have bordered pits, and also sometimes wall thickenings (Fig. 1.7). Wall thickenings may be annular (in a series of rings), helical (spiral) or arranged in a scalariform or reticulate mesh. Annular and helical thickenings are the types most commonly found in the first-formed (protoxylem) elements. The later-formed primary tracheary elements (metaxylem) and also the tracheary elements of the secondary xylem typically have bordered pits in their walls. These vary considerably in size, shape and arrangement. They may be oval, polygonal or elongated (scalariform), and organised in transverse rows (opposite pitting) or in tightly packed arrangement (alternate pitting). An evolutionary series from tracheids to vessel elements is widely recognised, due to the preponderance of tracheids in lower vascular plants and vessel elements in angiosperms, and many other trends of specialisation are widely established (Metcalfe and Chalk, 1983).

Phloem is the food-conducting tissue. The conducting cells of the phloem are called sieve elements, and are linked axially to form sieve tubes. Sieve elements have thin walls with characteristic sieve areas,

Fig. 1.9 *Deutzia maliflora.* Cross section of woody stem in region of vascular cambium, with secondary phloem (above) and secondary xylem (below). c = vascular cambium, r = ray, s = sieve element, v = vessel element.

which consist of groups of pores and associated callose. Sieve areas connect adjacent sieve elements. Sieve elements may be either sieve cells (most commonly found in lower vascular plants), or sieve tube elements, which are more specialised cells found in most angiosperms. In sieve cells the sieve areas are distributed throughout the cell wall, but in sieve tube elements they are mainly localised on the adjoining end walls, in the form of sieve plates. Sieve plates may be simple or compound. Mature sieve elements retain their contents at maturity, but usually lack nuclei. Behnke (1972, 1975, 1981) has shown that differences in the ultrastructure of sieve tube plastids have important systematic and evolutionary applications. Sieve tube elements commonly have associated specialised parenchyma cells, termed companion cells, which are derived from the same initial in the vascular cambium as the adjacent sieve tube element.

2

The stem

2.1 Shoot apex

Many authors have attempted to define conveniently the recognisable zones of activity which partition the shoot apical meristem of angiosperms. The most widely accepted theory is still that proposed by Schmidt (1924), which divided the central apical zone into two main regions: tunica and corpus. This provides a useful framework for descriptive studies, but requires a flexible application. The tunica, which varies in thickness, represents the outer (one to six, but most commonly two) layers of cells, with cell divisions mainly anticlinal (at right angles to the surface). The corpus is the region proximal to the tunica, with cell divisions oriented in all directions. The difference between the two layers is largely quantitative, and there is often a slight intergradation between them, the outermost corpus layers having more anticlinal cell divisions than the layers within them. There may be variation in size and distinctness of zonation even within different stages of development of the same plant. As Clowes (1961) pointed out, a consideration of the mechanical aspects of apical growth leads to an expectation of a system in which the outer layers contribute to surface growth and the inner layers to an increase in volume.

The central apical cells of both tunica and corpus layers are sometimes relatively larger and more highly vacuolated than those on either side, and are termed tunica or corpus initials. The central region underlying the corpus layer is a rib meristem giving rise to files of cells which later become the pith. This central region is surrounded by a peripheral flank meristem which produces the procambium, cortical region and leaf primordia (Fig. 2.1).

The vegetative shoot apex contributes to extension growth of the

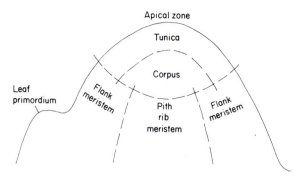

Fig. 2.1 Diagram of angiosperm shoot apex organisation. (Adapted from Gifford and Corson, 1971.)

shoot and initiates leaf primordia. However, some shoot apices become determinate and cease terminal growth, for example in the development of shoot thorns, although these may sometimes revert to vegetative growth. Reproductive shoot apices are more complex examples of determinate growth. The reproductive apex differs from the vegetative apex in size and shape and areas of mitotic activity. In general, at floral transition there is an overall increase in mitotic activity at the apex, but a proportionally greater increase among the axial apical cells than the peripheral cells (Gifford and Corson, 1971).

2.2 Primary stem structure

The plant stem is generally cylindrical, or less commonly ridged or quadrangular (Figs 1.1, 2.2). Primary vascular tissue usually consists of either a complete cylinder or a system of discrete vascular bundles. The cortex is the region of ground tissue between the vascular tissue and the epidermis, and the junction between the cortex and vascular region is often called the pericyclic region, from which endogenous adventitious roots may arise (section 3.3). Where there is a distinct central region of ground tissue, this is termed the pith, although in many stems the pith breaks down to form a central hollow cavity.

The stem epidermis, usually a single layer of cells, may often have stomata and trichomes, as in the leaf epidermis (section 4.2). The stem primary ground tissue is basically parenchymatous but may be modified into various tissue types or interspersed with fibres and sclereids.

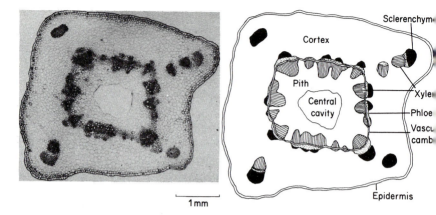

Fig. 2.2 *Vicia faba* (broad bean). Cross section of stem.

Parenchyma cells frequently become lignified as the plant ages. Square or angled stems often have strengthening collenchyma at the angles, immediately within the epidermis. Many stems are photosynthetic organs with a chlorenchymatous cortex, particularly in leafless (apophyllous) plants, such as many Juncaceae. Leafless plants often occur in dry, or xeric, environments, and frequently have xeromorphic modifications such as sunken stomata or large areas of sclerenchyma (Böcher and Lyshede, 1968, 1972) (section 4.5). Some plant stems have secretory cells or ducts in the ground tissue. For example, many species of *Euphorbia* have branched networks of laticifers in the cortex, which often extend throughout the ground tissue of the stem and leaves (Fig. 2.3).

Plants with succulent stems, such as many Cactaceae, typically have areas of large thin-walled cells which contain a high proportion of water. Many stems store food reserves in the form of starch granules, most commonly in the inner cortex, but also frequently throughout the rest of the ground tissue. Starch is most abundant in stems that are specialised as storage or perennating organs, such as corms of *Crocus* and other seasonally active plants. Sometimes the layer of cortical cells immediately adjacent to the vascular tissue is distinct from the rest of the cortex, and packed with starch granules. It is termed a starch sheath, or sometimes an endodermoid layer or endodermis, although the component cells usually lack the Casparian thickenings typically found in the root endodermis (section 3.2.2).

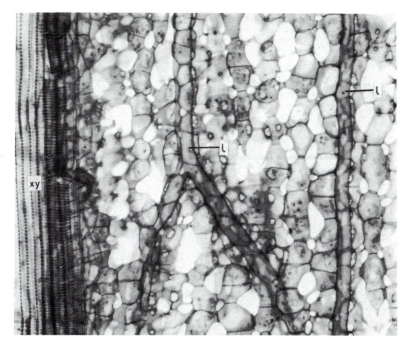

Fig. 2.3 *Euphorbia balsamifera*. Longitudinal section of pith of stem, showing laticifers (l) interspersed in ground parenchyma; xy = xylem. (x 150)

2.2.1 Primary vascular system

The primary vascular system derives from the procambium near the shoot apex. Vascular bundles may be collateral, with xylem and phloem adjacent to each other (Fig. 2.2), bicollateral, with phloem on both sides of the xylem, or amphivasal, with xylem surrounding the phloem.

In dicotyledons the vasculature of the internodal regions of the stem is typically arranged either in a continuous cylinder, or in a cylinder of separate or fused collateral bundles, with the phloem external to the xylem (Fig. 2.2). In some stems the bundles may be bicollateral; for example in species of *Cucurbita* internal phloem is present in addition to the external phloem. Amphivasal bundles are relatively unusual in dicotyledon stems. The vascular cambium, which produces secondary vascular tissue, is normally situated between the xylem and phloem, eventually forming a complete cylinder. Some stems also have cortical or medullary (pith) bundles, which may be associated with the leaf vasculature.

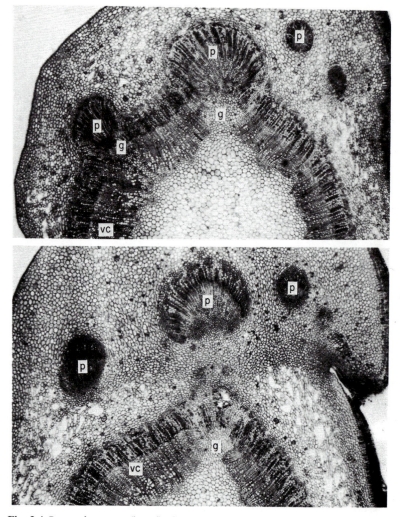

Fig. 2.4 *Prunus lusitanica* (laurel). Cross section of part of stem at node, showing connection of petiole vasculature (p) to main vascular cylinder (vc) of stem, with leaf gaps (g). (x 35)

The nodal anatomy of dicotyledons is often characteristic of families and therefore taxonomically significant, particularly the number and arrangement of leaf traces and gaps. Leaf and stem vasculature are connected at the node, with gaps in the stem vascular cylinder beneath their point of contact (Fig. 2.4). Nodes may be unilacunar, trilacunar or

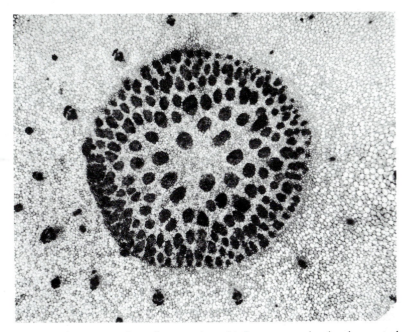

Fig. 2.5 *Libertia paniculata*. Cross section of inflorescence axis, showing central vascular region with numerous distinct vascular bundles, and outer cortex with few vascular bundles. (x 15).

multilacunar, depending on the number of gaps (lacunae) in the main vascular cylinder. This feature is clearer in stems where the vascular cylinder is complete, especially where a limited amount of secondary thickening has taken place; as a result, nodal anatomy has been studied far more extensively in woody than herbaceous plants (Howard 1974, 1979). The number of leaf traces departing from each gap is also variable; for example in *Clerodendrum* two traces depart from a single gap, and in *Prunus* a single trace departs from each of three gaps in the central vascular cylinder (Fig. 2.4). Nodal vasculature is further complicated by the axillary bud traces, which are connected to the main stem vasculature immediately above the leaf gaps. In the majority of cases there are two traces to each bud or branch. Shigo (1985) showed that our understanding of how tree branches are attached to trunks is often imperfect, although relevant for pruning techniques, and to the spread of tree diseases. The vascular tissue at the base of the branch is oriented abruptly downwards and forms a collar around the branch base. This branch collar is enveloped by a trunk collar, which links the vascular tissue of the trunk above and below the branch. There is no direct connection of xylem from the trunk above a branch into the branch xylem, as the tissues are oriented at right angles to each other. If a branch dies, a protection zone forms around its base to prevent a spread of infection into the trunk, and the branch is often shed.

In cross sections of monocotyledonous stems (Fig. 2.5) the vascular bundles are often scattered randomly throughout the central ground tissue, or sometimes arranged in two or more distinct rings. Bundles may be collateral, bicollateral, or most commonly amphivasal (especially in rhizomes), and lack a vascular cambium. Cortex and pith are sometimes indistinct, although the cortex may be defined by an endodermoid layer, or a distinct ring of vascular bundles, or in some stems, particularly inflorescence axes, by a cylinder of sclerenchyma enclosing the majority of vascular bundles. In a series of investigations using cinematographic techniques of analysis, Zimmerman and Tomlinson (1965, 1966, 1972) showed that the vascular system of monocotyledons is often extremely complex. Each major bundle, when traced on an upward course from any point in the stem, may have several bridges or branches before passing into a leaf. One of its major branches then continues a similar upward course towards the apex. In some palms there may be literally thousands of vascular bundles in a single cross section of the stem, although in most monocotyledons the number is much lower.

2.3 Stem growth in thickness

Increase in height, achieved by growth at the apical meristem, is often accompanied by a corresponding increase in stem thickness. This is brought about by means of different meristems in dicotyledons and monocotyledons.

2.3.1 Secondary thickening in dicotyledons

In dicotyledons, secondary vascular tissue (both xylem and phloem) is produced by the vascular cambium (Fig. 2.6), which usually becomes active at a short distance behind the stem apex. The amount of secondary vascular tissue produced is extremely variable, depending on the habit of the plant; in a few herbaceous dicotyledons (e.g. species of *Ranunculus*) it is completely absent. The vascular cambium generates

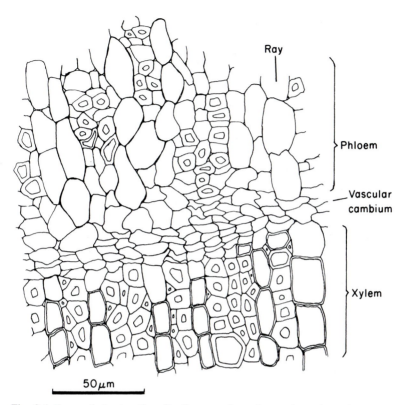

Fig. 2.6 *Prunus lusitanica* (laurel). Cross section of stem in region of vascular cambium.

secondary xylem (wood) at its inner edge (centripetally) and secondary phloem at its outer edge (centrifugally), although plants with anomalous secondary growth do not always follow this pattern.

The vascular cambium is a complex tissue consisting of both fusiform initials and ray initials, which form the axial and radial systems respectively. Fusiform initials are axially elongated cells with tapering ends (Fig. 2.7). They divide periclinally to form the axial elements of secondary tissues: tracheary elements, fibres and axial parenchyma in secondary xylem, and sieve elements, companion cells and fibres in secondary phloem. Ray initials are more or less isodiametric cells that divide periclinally to form ray parenchyma cells in both xylem and phloem. Fusiform initials sometimes give rise to new ray initials as the stem increases in circumference, and new rays are formed.

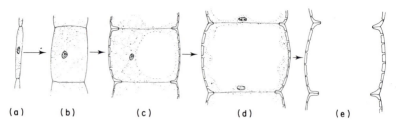

Fig. 2.7 Ontogeny of a vessel element (e) from a fusiform initial (a) in *Robinia pseudoacacia*. (After Eames and MacDaniels, 1925.)

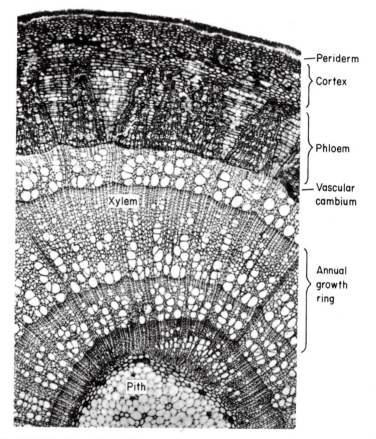

Fig. 2.8 *Tilia olivieri* (lime). Cross section of twig of just over three years growth. (x 25)

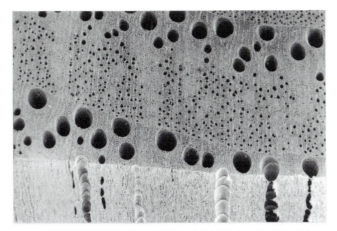

Fig. 2.9 *Quercus robur* (oak). Block of wood at edge of transverse and tangential longitudinal surfaces, showing large early (spring) wood vessels, and small late wood vessels. (SEM, x 40)

Secondary xylem of woody dicotyledons (known commercially as hardwoods) varies considerably in texture, density and other properties, depending on the size, shape and arrangement of the constituent cells (Metcalfe and Chalk, 1983). Wood is composed of a matrix of cells (Figs 2.8, 2.10), some arranged parallel to the long axis (fibres, vessels and chains of axial parenchyma cells), and others (ray parenchyma cells) forming the wood rays that extend radially from the vascular cambium towards the pith. The variation between different types of wood ensures that many woods can be fairly accurately identified by their anatomical structure (Miles, 1978; Schweingruber, 1990). To observe their structure, woods are cut in cross section and two longitudinal planes of section: along the rays (radial longitudinal section) and at right angles to the rays (tangential longitudinal section). In *Quercus* (oak) the vessels are solitary in cross section (Figs 2.9, 2.10), but in other woods they may be in clusters or radial chains. Axial parenchyma cells may be independent of the vessels (apotracheal) or associated with them (paratracheal), and are sometimes in regular tangential bands. Relative abundance of axial parenchyma varies, from sparse (or even completely absent) to rare cases such as *Ochroma pyramidale* (balsa) wood, where axial parenchyma cells are often more abundant than fibres. Variation in ray structure is best observed in longitudinal (particularly tangential) sections. Rays are termed uniseriate if they are one cell wide tangentially,

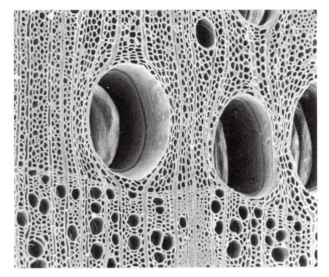

Fig. 2.10 *Quercus robur* (oak). Cross section of wood, showing growth ring boundary, with large solitary vessels in early (spring) wood, small solitary vessels in late wood, and uniseriate rays. (SEM, x 150)

and multiseriate if they are more than one cell wide. Sometimes both uniseriate and multiseriate rays occur in the same wood, for example in *Quercus* (oak). Ray cells vary in shape; homocellular rays are composed of cells of similar shapes, whereas in heterocellular rays the cells are of different shapes.

Other aspects of variation in the structure of hardwoods include the occurrence in some woods of either axial or radial secretory canals, or the storied (stratified) appearance of various elements, particularly rays, or the fine structure of the vessel walls (intervascular pitting, perforation plates and wall thickenings: section 1.5). For example, in some woods, such as *Prunus* spp. (cherry) or *Tilia cordata* (lime: Fig. 1.7), the vessel element walls are helically thickened, and in other woods (e.g. many Leguminosae) the pit apertures are surrounded by numerous warty protruberances, known as vesturing. Perforated ray cells, an unusual and interesting feature of some woods, are ray cells that link two vessel elements and themselves resemble and function as vessel elements, with perforation plates corresponding to those of the adjacent vessel elements. They occur sporadically in some woods, and serve as an example of the adaptable nature of the vascular cambium, since they are formed from ray initials.

Many woody temperate plants have seasonal (usually annual) cambial activity, which results in the formation of growth rings. The secondary xylem formed in the early part of the season (early wood or spring wood) is generally less dense and consists of thinner-walled cells than the xylem formed later in the growing season (late wood or summer wood). In ring porous woods the vessels are considerably larger in early than in late wood (Fig. 2.9), although in diffuse porous woods the main distinction is in size and wall thickness of the fibres. As woody plants age and their trunks increase in girth, the central area becomes non-functional with respect to water transport or food storage, and often the vessels become blocked by tyloses. Tyloses are formed when adjacent parenchyma cells grow into the vessels through common pit fields. The central non-functional area of the trunk, the heartwood, is generally darker than the outer living sapwood.

There are may examples of woody dicotyledons, particularly climbing plants (lianes), in which secondary growth does not fit the "normal" pattern of xylem and phloem production, and is termed "anomalous". These anomalous forms occur in many plant families. For example, some plants have areas of phloem (included or interxylary phloem) embedded in the xylem, either in islands (e.g. in *Avicennia*), or in alternating concentric bands. Other examples may have irregularly divided or deeply fissured areas of xylem and phloem, a flattened stem, a deeply divided and irregularly shaped stem, or a quadrangular stem (Fig. 2.11) with the xylem deeply furrowed by regular areas of xylem (Schenck, 1893; Obaton, 1960; Metcalfe and Chalk, 1983). Anomalous forms are achieved either by the formation of new vascular cambia in unusual positions, or by the unusual behaviour of the existing cambium, in producing phloem instead of xylem at certain points.

Secondary phloem in dicotyledons, also a product of the vascular cambium, similarly displays great variation in structure. However, this variation has been the subject of far fewer studies than secondary xylem, and is less readily used as a means of identification. As in secondary xylem, secondary phloem consists of both axial and radial systems, formed from the fusiform and ray initials respectively. Phloem rays are continuous with xylem rays, and may be similarly uniseriate or multiseriate, although they are often dilated towards the cortex as a result of cell divisions to accommodate increase in stem thickness (Fig. 2.8). The parenchymatous ray cells are often difficult to distinguish from cortical cells at their junction, and older ray cells sometimes become lignified to form sclereids. The axial system of the phloem consists of

Fig. 2.11 *Tynanthus elegans* (a liane, or climbing plant, of the family Bignoniaceae). Cross section of woody stem showing anomalous secondary growth: xylem region with four deep fissures of phloem. (x 10)

sieve elements and companion cells, as in primary phloem (section 1.5), and also often includes fibres, sclereids and axial parenchyma cells. In some cases, such as *Tilia* (lime: Fig. 2.8), the fibres are formed in groups at regular intervals, resulting in characteristic tangential bands of fibres alternating with areas of sieve elements and parenchyma cells.

2.3.2 Stem thickening in monocotyledons

In many herbaceous monocotyledons there is little or no stem thickening growth, but where increase in stem thickness does occur, it is initiated at the primary thickening meristem near the vegetative shoot apex (Fig. 2.12). In some monocotyledons further increase in thickness also occurs at a greater distance from the apex, either by diffuse secondary growth, or at the secondary thickening meristem (Fig. 2.13).

Most monocotyledons, especially those with short internodes and crowded leaves, have a primary thickening meristem. This is situated in the pericyclic region just below the apex. It consists of a relatively narrow zone of meristematic cells producing radial derivatives, usually parenchyma towards the outside (centrifugally), and both parenchyma

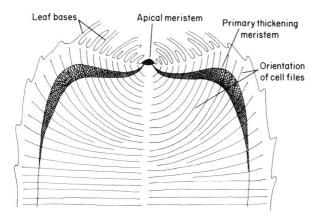

Fig. 2.12 Diagram of longitudinal section of the crown of a typical thick-stemmed monocotyledon with a primary thickening meristem, showing orientation and extent of radial derivatives. Vascular strands not shown. (Adapted from DeMason, 1983).

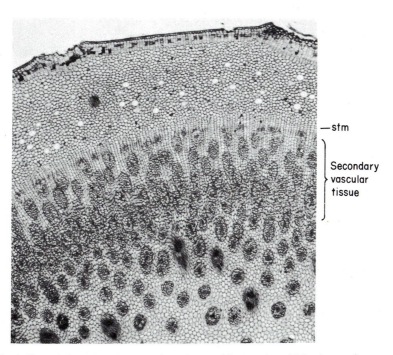

Fig. 2.13 *Cordyline rubra*. Cross section of stem with secondary thickening meristem (stm). (x 20)

and discrete vascular bundles towards the inside (centripetally). The primary thickening meristem is responsible for primary stem thickening, adventitious root production, and formation of linkages between root, stem and leaf vasculature.

In many monocotyledons the primary thickening meristem ceases activity at a short distance behind the apex, and subsequent stem thickening is limited. In palms, however, considerable stem thickening occurs by extensive division and enlargement of ground parenchyma cells. This is termed diffuse secondary growth. In some woody Liliiflorae, such as species of *Aloe*, *Yucca* and *Agave*, further increase in stem thickness is achieved by secondary growth from a secondary thickening meristem.

The secondary thickening meristem is essentially similar to the primary thickening meristem in that it is situated in the pericyclic region and produces similar radial derivatives. However, the two meristems differ in that the secondary thickening meristem occurs further from the stem apex, and the secondary vascular bundles produced are often amphivasal and radially elongated, although there are no precise criteria for distinguishing between derivatives of the two meristems, and transitional forms exist. In some species (such as *Beaucarnea recurvata* and *Cordyline terminalis*) the primary and secondary thickening meristems are often longitudinally discontinuous (Stevenson, 1980; Stevenson and Fisher, 1980), whereas in others (such as *Yucca whipplei*) they have been shown to be longitudinally continuous (Diggle and DeMason, 1983). The close similarity between the two meristems has led some authors to regard them as developmental phases of the same meristem (Rudall, 1991).

The primary and secondary thickening meristems of monocotyledons are not homologous with the vascular cambium of dicotyledons because the vascular derivatives are arranged in completely different ways. In dicotyledons the vascular cambium produces phloem centrifugally and xylem centripetally, whereas in monocotyledons, discrete vascular bundles of both xylem and phloem are formed centripetally in a parenchymatous ground tissue. Furthermore, the primary thickening meristem, which originates in ground tissue, is a tiered meristem, often fairly diffuse (especially near the shoot apex: Fig. 2.12), whereas the vascular cambium of dicotyledons is uniseriate, and initially originates within vascular tissue.

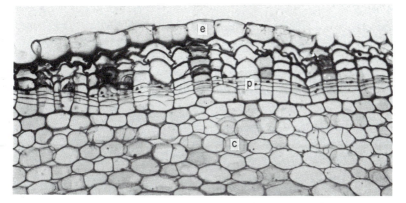

Fig. 2.14 *Sambucus nigra* (elder). Cross section of stem surface, showing broken epidermis (e), with periderm (p) forming beneath, in outer cortical layers; c = cortex. (x 130)

2.4 Periderm

Periderm is a protective tissue of corky (suberinised) cells that is often produced as a response to wounding, or in the stem or root cortex. It consists of up to three layers: phellogen, phellem and phelloderm. The phellogen is a uniseriate meristematic layer of thin-walled cells which produces the other two layers, the phellem to the outside, and (in some cases) the phelloderm to the inside. The phellem cells constitute the corky tissue. They are tightly packed cells devoid of contents at maturity. They have deposits of suberin and sometimes lignin in their walls, and form an impervious layer to prevent water loss and protect against injury. Phelloderm cells are non-suberinised and paren-chymatous, and contribute to the secondary cortex.

A periderm commonly occurs in the cortex of secondarily thickened stems, to replace the epidermis, which splits and peels away (Fig. 2.14). The phellogen may originate either adjacent to the epidermis (or even within the epidermis) or deeper in the cortex. Sometimes several phellogens may form almost simultaneously. The pattern of periderm formation largely dictates the appearance of the bark of a woody plant (the term "bark" usually denoting all the tissues outside the vascular cambium). For example, the smooth papery bark of a young birch tree is formed because the periderm initially expands tangentially with the increase in stem diameter, but later flakes off in thin papery sheets as abscission bands of thin-walled cells are formed. In the trunk of *Quercus suber* (cork oak), the source of commercial cork, the initial phellogen may continue activity indefinitely, and produces seasonal growth rings, although in the commercial process it is removed after about 20 years to make way for a second, more vigorous phellogen, which provides the commercial cork. Many barks have lenticels, which are areas of loose cells in the periderm, often initially formed beneath stomata in the epidermis, and thought to be similarly concerned with gaseous exchange.

3

The root

3.1 Root apex

Root apices have a terminal protective root cap composed of several layers of parenchymatous cells. The root apical meristem is adjacent and proximal to the root cap, and the junction between the cap region and the meristem may either be clearly defined by a distinct cell boundary ("closed" structure, as in *Zea mays*, Fig. 3.1), or ill-defined

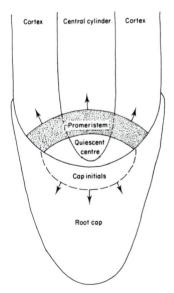

Fig. 3.1 Diagram of organisation of root apex in *Zea mays* (maize), a species with "closed" structure. Arrows indicate direction of displacement of cell derivatives. (Adapted from Feldman, 1984.)

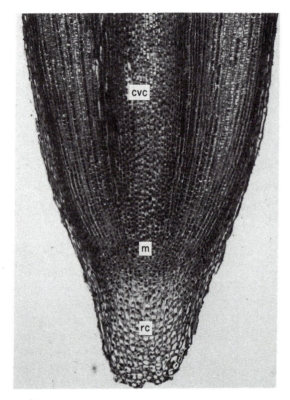

Fig. 3.2 *Vicia faba* (broad bean). Longitudinal section of root apex; "open" structure. cvc = central vascular cylinder, m = meristem, rc = root cap. (x 60)

("open" structure, as in *Vicia faba*, Fig. 3.2). In some taxa distinct meristematic layers can be identified, for example in *Zea mays* one layer gives rise to the root cap, another to the epidermis, and the innermost layer to the vascular tissue. However in other taxa, such as *Vicia faba* (Fig. 3.2) there is an undifferentiated common initiating region for all root tissues (Steeves and Sussex, 1989). Views on the organisation of the root apical meristem have in the past differed widely (Clowes, 1961). However, the discovery of the "quiescent centre", a group of relatively inactive cells at the very centre and tip of the apical meristem (Fig. 3.1), has led to an adjustment in our understanding of the zonation of mitotic activity. Contemporary opinions suggest that most cell division activity occurs in the area of cells proximal to the quiescent centre, producing the bulk of the root tissue. This active region is termed the promeristem. The cells of the quiescent centre divide infrequently, and its role is

somewhat obscure, although many investigators believe that it influences the cellular organisation of the root (Feldman, 1984). The
quiescent centre is defined only by cell division activity, and is not
always distinct in root sections.

The cells of the root cap are also initially derived from the apical
meristem. However, ontogenetic studies on *Zea mays*, a species with
"closed" root apical structure (Fig. 3.1), have shown that the cap
initials become established and independent from the apical meristem at
an early stage in seedling development (Barlow, 1975). The cap
meristematic cells, located adjacent (distal) to the quiescent centre,
produce derivatives that are eventually displaced towards the outside of
the root cap, and subsequently sloughed off, contributing to the external
slime. Cells are generated and lost in the root cap at approximately the
same rate.

3.2 Primary root structure

Although roots can originate from various organs in the plant (section
3.3), basic primary structure is relatively uniform, and different to that
of the stem. Each root has an epidermis, cortex and central vascular
region (Fig. 3.3). The central region is usually bounded by an
endodermis and pericycle.

3.2.1 Root epidermis

The root epidermis is typically uniseriate, as in other parts of the plant.
In underground roots it often becomes worn away, and is replaced as an
outer layer either by a periderm (in most woody dicotyledons), or an
exodermis. In a few monocotyledons, particularly those with aerial
roots, such as some tropical Orchidaceae and Araceae, the root
epidermis is multiseriate, termed a velamen, which is commonly believed
to have a function in water absorption, by retaining water after rain.

Most angiosperms have root hairs, formed by extension of epidermal
cells. In some plants only certain root epidermal cells, the trichoblasts
(the smaller product of an unequal cell division: section 1.4.3), are
capable of root hair production. Root hairs usually arise about a
centimetre from the root apex, beyond the meristematic region, but in
an area where cells are still enlarging. In general they persist for only a
limited period before withering. This region of the root is the most

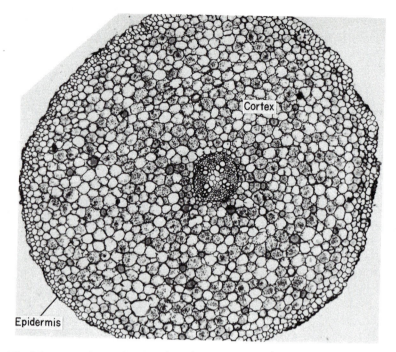

Fig. 3.3 *Ranunculus* sp. Cross section of root.

active in absorption of water, and the root hairs serve to present a greater surface area for this purpose.

3.2.2 Cortex

The cortex is the region between the pericycle and the epidermis, including the outer exodermis (where present) and the innermost layer, the endodermis, which is derived from cortical cells. Apart from these specialised layers, most cortical cells are parenchymatous and often perform an important storage function. In some plants, such as *Daucus carota* (carrot), the tap root is a modified swollen storage organ with a wide cortex. In most roots the bulk of the cortical cells are formed by sequential periclinal divisions, the innermost cells (later the endodermis) being the last formed.

Many plants with underground stems (corms, bulbs or rhizomes), particularly bulbous or cormous monocotyledons such as *Crocus*, *Freesia* or *Hyacinthus*, periodically produce contractile roots which

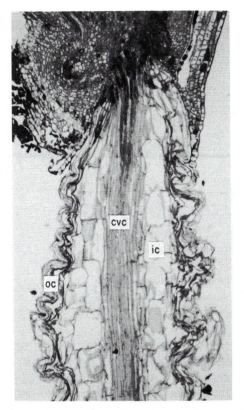

Fig. 3.4 *Tigridia dugesii.* Longitudinal section of contractile root at base of bulb, showing central vascular cylinder (cvc), expanded inner cortical cells (ic) and collapsed outer cortical cells (oc). (x 45)

draw the stem deeper into the soil (Fig. 3.4). These roots grow downwards, and then shorten vertically and expand radially (Jernstedt, 1984). They are recognisable by their wrinkled surface, and characteristically have areas of collapsed outer cortical cells, and occasional enlarged, thicker-walled middle and outer cortical cells (Fig. 3.4). The actual mechanism of root contraction has been the subject of some debate. Some investigators, such as Ruzin (1979) and Wilson and Anderson (1979), have found that the contraction process is initiated by active cell enlargement (radial growth, although accompanied by some axial shortening) in the inner cortex, followed by collapse of outer cortical cells, and subsequent surface wrinkling. However, other investigators, such as Sterling (1972), have suggested that the collapse of outer cortical cells results from the difference between atmospheric pressure and relatively low xylem pressure (due to transpiration), causing centripetal loss in turgidity.

The exodermis, a tissue that occurs sometimes in monocotyledon roots (e.g. in *Iris*), is formed from a few subepidermal layers of cortical cells that become suberinised or lignified as the epidermis is worn away in older roots.

The endodermis, which is a characteristic feature of roots, is a uniseriate cylinder of cortical cells surrounding the central vascular region, adjacent to the pericyle. These cells have become modified, typically by deposition of a band of suberin or lignin (Casparian strip) in their primary walls, forming a barrier against non-selective passage of water through the endodermis. Older endodermal cells often have thick lamellated secondary walls, in most cases on the inner periclinal walls, so that the Casparian strip is not apparent. The secondary wall deposits are often lignified, and also serve as an effective barrier to water. Occasional endodermal cells (passage cells) may remain thin walled, probably for passage of water between the cortex and vascular region.

3.2.3 Central vascular cylinder

The vascular tissue in the centre of the root is surrounded by one (or rarely more) layers of thin-walled cells, which constitute the pericycle (Fig. 3.5). The pericycle is potentially meristematic in younger roots, as it is the site of lateral root initiation, but in older roots it is less active, and may become lignified. The primary vascular tissue consists of several strands of phloem alternating with the "arms" (archs) of a central area of xylem which appears star-shaped in cross section. The protoxylem elements, which are the first-formed and narrowest in diameter, are at the tips of the "arms", nearest to the pericycle, with the wider metaxylem elements closer to the centre of the root. Similarly, as both xylem and phloem are exarch in the root (i.e. they mature centripetally), the protophloem is closer than the metaphloem to the pericycle. Roots may have two, three, four or more protoxylem poles, in which case they are diarch, triarch, tetrarch or polyarch respectively. However, there is often variation in the number of xylem poles, even sometimes within the same plant, depending on the diameter of the root. Dicotyledonous roots usually have a relatively low number of xylem poles, most commonly two, three or four, but many monocotyledons have a much higher number. In a few polyarch monocotyledons, such as

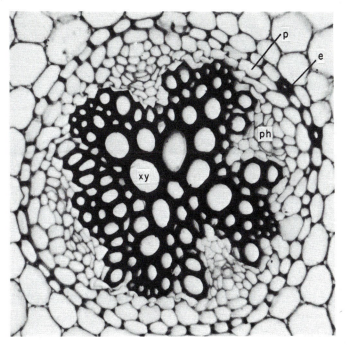

Fig. 3.5 *Ranunculus acris* (buttercup). Cross section of central part of root without secondary thickening. Xylem (xy) with four protoxylem poles; four areas of phloem (ph); pericycle (p); endodermis (e). (x 380)

Iris, the centre of the root is parenchymatous, sometimes becoming lignified in older roots. More commonly the central region is occupied by a group or ring of xylem vessels.

3.3 Initiation of lateral and adventitious roots

Lateral roots are branches of the tap root, and have a deep-seated (endogenous) origin. Although the most recently formed lateral roots are usually those nearest to the root apical meristem, they are initiated in relatively mature tissues some distance from the apex, often in acropetal sequence. In angiosperms the initial cell divisions in lateral root formation usually occur in the pericycle. This forms a lateral root primordium, with in many cases some subsequent cell divisions in the endodermis, so that ultimately both the pericycle and the endodermis contribute to the tissues of the lateral root. The growing lateral root

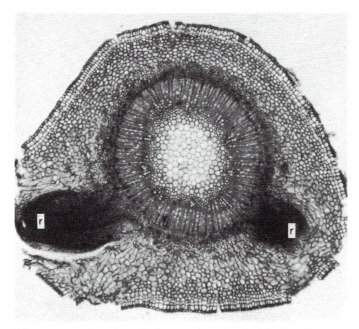

Fig. 3.6 *Ligustrum vulgare* (privet). Cross section of stem with adventitious roots (r). (x 40)

pushes its way through the cortex and epidermis of the parent root, either by mechanical or enzymatic action. In dicotyledons the position of lateral root initiation in the pericycle is usually at a point adjacent to a protoxylem pole, unless the root is diarch, in which case initiation is sometimes opposite a phloem pole. In monocotyledons, lateral root initiation can be opposite either protoxylem or phloem poles, although in roots with a large number of vascular poles it is often difficult to determine the precise site of initiation (McCully, 1975).

The term adventitious root, used in its broadest sense, refers to any root not arising from the seedling radicle or its branches (Esau, 1965). Adventitious roots may be deep-seated (endogenous) in origin (Fig. 3.6), or more rarely exogenous, arising from superficial tissues such as the epidermis (e.g. in surface-rooting *Begonia* leaves). In most monocotyledons, adventitious roots arise from cell divisions in the pericyclic region of the stem; the primary thickening meristem contributes to adventitious root formation (section 2.3.2). Adventitious roots are often formed at nodes on the stem, which is why cuttings are commonly taken from just below a node.

3.3.1 Roots associated with micro-organisms

Many vascular plants form symbiotic relationships with soil micro-organisms. In legumes, nitrogen-fixing bacteria invade the root cortex through root hairs, and stimulate meristematic activity in the cortex (and sometimes also in the pericycle) to form a root nodule, which often becomes elongated (Steeves and Sussex, 1989) to resemble a short lateral root. Other soil micro-organisms may induce the formation of modified lateral roots. For example, in many woody angiosperms, invading filamentous bacteria promote the development of short, swollen lateral roots, and in some temperate woody forest species, especially in the families Fagaceae and Betulaceae, ectomycorrhizal fungi form a mantle over stunted lateral roots. On the other hand, the more common endomycorrhizal fungi, which invade the cells of the host root, often have little influence on root morphology.

3.4 Secondary growth in roots

In monocotyledons, secondary growth in roots is extremely rare, even among species with a secondary thickening meristem (section 2.3.2). Among arborescent monocotyledons it is confined to roots of *Dracaena* (Tomlinson and Zimmermann, 1969), and even here it is somewhat limited. It initiates in the pericycle or cortex, or sometimes both, and the secondary vascular tissue is similar to that of the stem.

Most dicotyledonous roots have a certain amount of secondary thickening (Fig. 3.8), with the exception of a few herbaceous species, such as *Ranunculus* (Figs 3.3, 3.5). In some tree species the thickening and strengthening of the root system may be important in supporting the trunk. As in the stem, the secondary vascular tissues of the root are produced by a vascular cambium. This initially develops in the regions between the primary xylem and phloem, then in derivatives of cell divisions in the pericycle next to the xylem poles. Since cambial activity proceeds in this sequence, the xylem cylinder soon appears circular in cross section (Fig. 3.7). Further pericyclic cell divisions result in the formation of a secondary cortex, and in many cases, particularly where secondary growth is extensive, the epidermis, primary cortex and endodermis become split and are sloughed off, and a periderm forms.

Root secondary xylem is usually fairly similar to that of the stem in the same plant, but may differ in several respects. For example, in ring

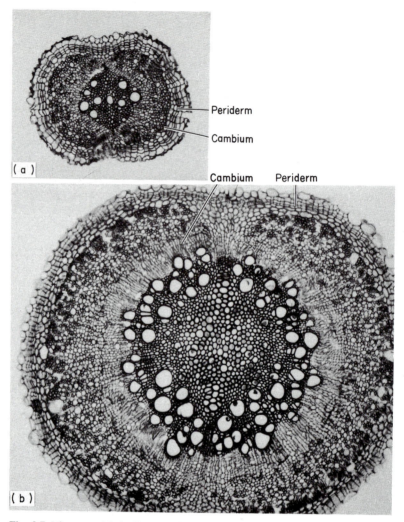

Periderm

Cambium

Cambium   Periderm

(a)

(b)

Fig. 3.7 *Ulmus* sp. (elm). Cross sections of roots with (a) secondary growth just begun, and (b) secondary thickening more extensive. (x 60)

porous species of *Quercus* (oak), such as *Q. robur*, the root wood is diffuse porous, the vessels being similar in size across each growth ring. This is in contrast to the stem, where the early wood vessels are markedly larger than the late wood vessels (Fig. 2.9). As with trunk wood, root wood of individual taxa often has identifiable characteristics (Cutler *et al.*, 1987).

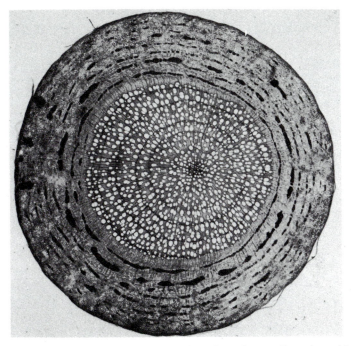

Fig. 3.8 *Populus tremula* (aspen). Cross section of woody root. Secondary phloem with concentrically arranged blocks of fibres; secondary xylem with four growth rings. (x 10)

3.5 Haustoria of parasitic angiosperms

There are many examples of angiosperms that are parasitic on the roots, stems and leaves of other angiosperms. Perhaps the best known are the mistletoes (e.g. *Viscum album*, family Viscaceae), but others include *Cuscuta* spp. (dodders, family Convolvulaceae), and members of the families Santalaceae (sandalwoods), Orobanchaceae (broomrapes) and Scrophulariaceae (parasitic figworts). The haustorium is a highly modified root through which nutrients pass from the host to the parasite. Haustoria may be termed primary or secondary; a primary haustorium is a direct outgrowth of the apex of the parasite radicle, whereas a secondary haustorium is a lateral organ which may develop from a modified adventitious root or from outgrowths of roots or stems (Kuijt, 1969). Haustoria take many different forms in different parasites. They often penetrate the host tissue to the xylem, and form a continuous

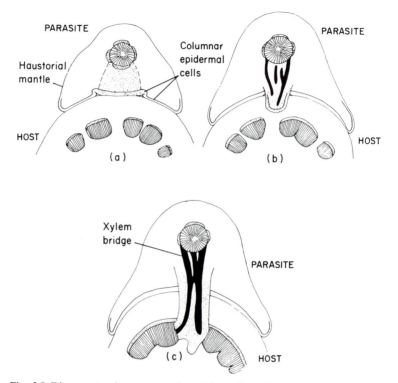

Fig. 3.9 Diagram to show penetration of host tissue by parasitic haustorium of *Cassythia* sp. (Adapted from Kuijt, 1969.)

xylem bridge between the host and the parasite. The phloem tissues are not linked in the same way, except sometimes by specialised parenchyma cells. In sandalwoods and many other parasites, the developing haustorium forms a mantle of mainly parenchymatous tissue around the host organ. The epidermal cells in contact with the host become elongated and secretory, and the centre of the haustorium develops an intrusive process which grows into the host by enzymatic and mechanical action (Fig. 3.9). In some Loranthaceae the haustorium does not invade the host tissues after forming a mantle, but instead influences the host tissue to form a placenta-like outgrowth of vascular tissue to supply nutrients to the parasite. When the parasite dies, a woody outgrowth of convoluted host tissue remains, known as a "woodrose".

4

The leaf

The mature lamina consists of an adaxial and abaxial epidermis enclosing several layers of mesophyll cells interspersed with a network of vascular bundles. Each tissue may be variously differentiated into specialised cell types, the degree of differentiation between them varying considerably between taxa. Angiosperm leaves display much diversity in cross-sectional outline. In dorsiventral or bifacial leaves the upper and lower surfaces differ, for example in relative numbers of stomata and trichomes, and the mesophyll may be differentiated into both palisade and spongy tissues (Fig. 4.1). In isobilateral or unifacial leaves the epidermis and mesophyll are usually more or less similar on both surfaces, and in centric or terete leaves (Fig. 4.9) they are continuous. Some monocotyledons, such as *Libertia* (Fig. 4.2) have unifacial leaves with a bifacial sheathing base.

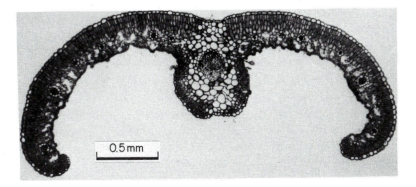

0.5 mm

Fig. 4.1 *Helichrysum italicum*. Leaf, cross section.

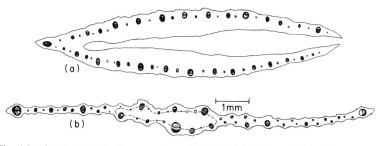

Fig. 4.2 *Libertia chilensis.* Cross sections of (a) bifacial leaf base, (b) isobilateral leaf blade.

4.1 Leaf development

Leaves are initiated as small conical projections (leaf primordia) close to the stem apex (Fig. 2.1), by periclinal divisions either in the outermost cell layers, or in the layers immediately below them. Elongation continues by apical growth for a short time in the primordia, then it is replaced by intercalary growth. In young primordia the adaxial marginal cells divide rapidly to form a flattened leaf blade. This marginal growth is suppressed in the region that later becomes the petiole, and in many monocotyledons it is often simultaneous with apical growth. Marginal growth is usually subsequently replaced by cell divisions across the whole leaf blade. By this stage the number of cell layers comprising the leaf thickness has been more or less established, and the whole lamina functions as a plate meristem. Cell divisions are mainly anticlinal, resulting in regular layers of cells which are disrupted only by the differentiation and maturation of the vascular bundles. Rates of growth and cell division may sometimes vary in different parts of the leaf. In many monocotyledons, meristematic activity governing leaf elongation is restricted to a region at the base of the leaf, the basal rib meristem, which results in axial files of cells of increasing maturity towards the distal end of the leaf (see also section 4.2.2).

Several investigators (e.g. Kaplan, 1970, 1973a,b, 1975, 1984) have attempted to relate differences in histogenesis in leaves to differences in mature leaf structure in various plant taxa. For example, the unifacial leaves of some monocotyledons with a bifacial sheathing leaf base and a unifacial upper blade (Fig. 4.2), have long been interpreted as resulting from suppressed marginal growth and emphasised adaxial growth. By means of comparative studies, both morphological and anatomical,

Arber (1925) and others have contended that the unifacial leaves of some monocotyledons are homologous with the phyllodes of some dicotyledons such as *Acacia*, and that both are derived from a petiolar structure, after suppression of the leaf blade. More recently, however, Kaplan (1975), by comparing the early development of several unifacial monocotyledon and dicotyledon leaves, accepted that the two structures are homologous, but as a result of an unusual method of leaf expansion by a single adaxial meristem rather than from petiolar derivation. In another investigation, Kaplan (1984) examined developmental mechanisms in other unrelated plants with similar structural features, such as dissected or lobed leaves in the families Araceae and Palmae, and found very different ontogenetic sequences. As a result of these and other investigations, the advantages of an ontogenetic approach to evolutionary problems are now widely recognised.

4.2 Leaf epidermis

The epidermis is a complex tissue that usually consists of a single layer of cells, although in rare cases (such as species of *Ficus* and *Peperomia*) it may have proliferated to form a multiple epidermis. This is similar in structure to a hypodermis (section 4.3), and can be distinguished from it only by a developmental study.

The specialised elements of the leaf epidermis are essentially the same as those of the stem: stomata, trichomes, papillae, surface sculpturing, epicuticular wax and variously differentiated epidermal cells. However, the leaf surface has been the subject of more investigations than other plant surfaces, and since many of the variable features are constant within taxa they often have taxonomic applications.

4.2.1 Epidermal cells

In surface view, epidermal cells may be elongated or more or less isodiametric, and the anticlinal walls (at right angles to the plant surface) may be straight (Fig. 4.3b) or undulating (Fig. 4.3a,c). Cells over veins are often elongated in the direction of the vein, and in long narrow leaves of the sort found in many monocotyledons, such as *Iris*, epidermal cells are frequently elongated along the long axis of the leaf. Anticlinal cell walls are often more sinuous on the abaxial than the adaxial surface of the same leaf.

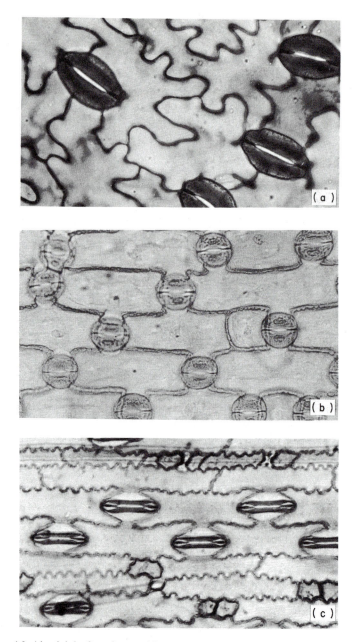

Fig. 4.3 Abaxial leaf surfaces with stomata. (a) *Paeonia officinalis*, (b) *Crocus speciosus*, (c) *Arundo donax*. (x 380)

Epidermal cells may also vary in size and wall thickness in different parts of the same leaf. A particularly striking example of this occurs in some Gramineae, such as *Zea mays*, where bulliform cells occur in restricted areas of the leaf epidermis. These cells are larger than other epidermal cells and thin walled, and may in some cases have a function in the unrolling of the leaf in response to turgor pressure and water availability.

Occasionally epidermal cells may contain crystals or silica bodies. Characteristic silica bodies are found in the leaf epidermis of many species, particularly in the monocotyledon families Cyperaceae (sedges), Gramineae (grasses) and Palmae (palms). In the Gramineae the epidermis typically consists of both long and short cells, the short cells sometimes forming the bases of hairs. Cystoliths are calcareous bodies sometimes found in cells of the epidermis (e.g. in the family Opiliaceae), or the underlying mesophyll (e.g. in *Ficus*), although even these may sometimes protrude above the leaf surface.

4.2.2 Stomata

Stomata are the pores in the epidermis through which gaseous exchange takes place. They occur on most of the plant surfaces above ground, especially on green photosynthetic stems and leaves, but also on floral parts. On leaves, stomata may be found on both surfaces (amphistomatic), or on the abaxial surface only (hypostomatic), or in rare cases, such as on floating leaves, on the adaxial surface only (epistomatic). They occur most frequently on the regions overlying the chlorenchymatous mesophyll rather than over the veins, and may sometimes be restricted to certain areas, such as in grooves or depressions (Fig. 4.13).

Each stoma consists of two kidney-shaped guard cells surrounding a central pore. Cuticular ridges may extend over or under the pore from the outer or inner edges of the adjacent guard cell walls. Stomata may be sunken (Fig. 4.4) or raised, and are often associated with a substomatal cavity in the mesophyll.

The epidermal cells immediately adjacent to the guard cells are termed subsidiary cells if they are different in shape to the surrounding epidermal cells. Classifications of stomatal types are based either on the arrangement of mature subsidiary cells, or on their patterns of development. The most widely used system of classification of mature stomatal types (summarised by Wilkinson, 1979) includes many terms such as: anomocytic (with no subsidiary cells), anisocytic (with three

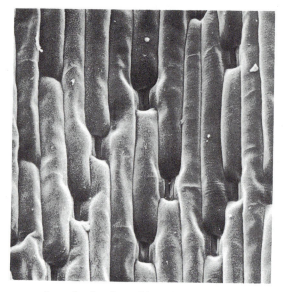

Fig. 4.4 *Crocus speciosus*. Leaf surface with sunken stomata. (SEM, x 400)

unequal subsidiary cells), diacytic (with one or more pairs of subsidiary cells with their common walls at right angles to the guard cells), and paracytic (with, at either side of the guard cells, one or more subsidiary cells which may or may not meet at the poles). However, this classification, although useful in a descriptive sense, has a limited application, as it does not take into account the different developmental pathways which may lead to similar stomatal types. It therefore tends to group stomatal types that may be non-homologous.

An alternative stomatal classification is based on developmental pathways. The importance of ontogenetic studies in comparing stomata of different taxa is now widely recognised. Stomatal ontogeny is particularly accessible in monocotyledons as their leaves grow with a basal rib meristem which often produces longitudinal files of epidermal cells with a sequence of development along the leaf (Tomlinson, 1974; Rasmussen, 1983). In monocotyledons there is an unequal mitotic division which produces a larger cell and a meristemoid (guard cell mother cell). If the development is agenous, the meristemoids give rise directly to the guard cells, and there are no subsidiary cells. In perigenous modes of development the meristemoid gives rise directly to the guard cells, and the subsidiary cells are formed from neighbouring cells, often by oblique divisions. In mesogenous modes of development the meristemoid undergoes a further mitotic division into two cells, of which one further subdivides to form the guard cells, and the other usually forms one or more subsidiary cells. Subsidiary cells derived from the meristemoid are termed mesogene cells, whereas those derived from the neighbouring cells are termed perigene cells. In some cases mesogene cells are not distinct from surrounding epidermal cells at maturity.

4.2.3 Trichomes and papillae

Trichomes are epidermal outgrowths that occur on all parts of the plant surface (Fig. 4.5). They vary widely in both form and function, and include unicellular or multicellular, branched or unbranched forms, and also scales, glandular (secretory) hairs, hooked hairs and stinging hairs. The distinction between trichomes (hairs) and papillae, which are also epidermal outgrowths, is not always clear, although papillae are generally smaller and unicellular. In cases where there are several small outgrowths on each epidermal cell, these outgrowths are usually termed papillae, but where there is only one unicellular outgrowth per cell, the distinction is dependent on size.

Trichomes may occur on the entire leaf surface, or may be restricted to certain areas, such as in grooves (Figs 4.5b, 4.13) or at the margins. Often there are several different types on the same leaf. For example, in many genera of Labiatae, such as *Hyptis* (Fig. 4.5), *Rosmarinus* (rosemary), *Thymus* (thyme) and *Lavandula* (lavender), there are two or more sizes of glandular hair, and branched or unbranched non-glandular hairs.

Glandular trichomes (Fig. 4.5a) usually have a one or more celled stalk and a secretory head with one to several cells. In some glandular hairs the substances secreted collect between the secretory cells and a raised cuticle, which may later break to release them. There are many different types of glandular hair, and they secrete a variety of substances, from salt or essential oils to digestive juices in some carnivorous plants (Fahn, 1979). The stinging hairs of *Urtica dioica* (stinging nettle) are rigid, hollow structures which contain a poisonous substance. The spherical tip of the hair is readily broken off in contact with an outside body, and the remaining sharp point may then penetrate the skin and release the fluid. Other examples of specialised hair types include the scales on the leaves of many Bromeliads, which are specially adapted for water absorption and the salt-secreting glands of species of *Avicennia*.

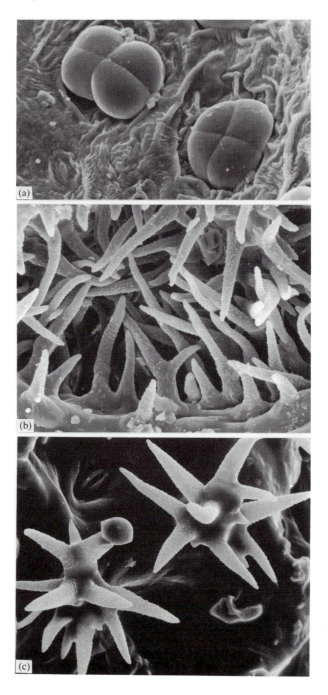

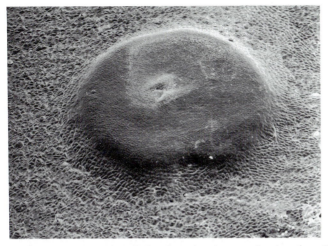

Fig 4.6 *Omphalea diandra.* Abaxial leaf surface, showing extrafloral nectary. (SEM, x 100)

4.2.4 Extrafloral nectaries

Apart from various types of glandular trichomes, some plants have specialised nectar-secreting regions (extrafloral nectaries) on the leaf or petiole. Although in some cases the function of the extrafloral nectaries remains in doubt, and may be related to regulation of surplus sugars, most are believed to have a role in insect – plant relations. For example, some (e.g. in *Acacia*) are known to attract ants, which protect the plant against potential insect herbivores. Extrafloral nectaries are often found over veins, or in the angles of principal veins, or at the proximal or distal ends of the petiole. As with floral nectaries (section 5.6), they may consist of groups of glandular trichomes (e.g. in some *Hibiscus* species), sometimes situated in specialised pockets (domatia), or they may be regions of anticlinally elongated secretory epidermal cells, often associated with underlying vascular tissue, and sometimes in pits or raised regions (e.g. in many Euphorbiaceae, such as *Aleurites*, *Hevea* and *Omphalea*: Fig. 4.6). Pearl glands, or pearl bodies, found in some

---

Fig. 4.5 Hairs on abaxial leaf surfaces of various *Hyptis* species (Labiatae) (SEM, x 500). (a) *H. caespitosa*, large sunken glandular hairs, with four-celled heads; (b) *H. proteoides*, surface depression lined with short non-glandular hairs of one to three cells, mainly unbranched, or rarely branched; (c) *H. emoryi*, branched hairs, the lower one with a small gland at the end of one branch.

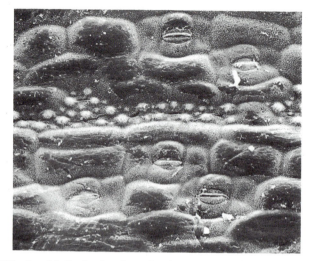

Fig. 4.7 *Gibasis pulchella.* Leaf surface, showing stomata, epicuticular wax particles and surface markings. (SEM, x 300)

dicotyledon families such as Leguminosae and Piperaceae (O'Dowd, 1982), are secretory, often globular, leaf emergences or trichomes that provide food (carbohydrates, lipids and proteins) for ants, which carry them off.

4.2.5 Cuticle and wax

The cuticle, which is composed mainly of cutin, lipids and waxes, is a layer covering the entire leaf surface (and most other aerial plant surfaces). In mesomorphic leaves the cuticle is often thin and almost transparent; however, many xeromorphic plants have thick cuticles which often appear lamellated in cross section. In some plants the outer surface has characteristic patterns of cuticular ridges or folds. These striations may be short or long and with regular or random orientation, sometimes radiating from around stomata or trichomes. Barthlott and Ehler (1977) examined epidermal features of many angiosperms using a scanning electron microscope, and found much variation in both cell shape and surface sculpturing, often with taxonomic applications. They discussed the biological significance of surface sculpturing in relation to mechanical and optical properties and wettability of the surface. Cutler and Brandham (1977) and Brandham and Cutler (1978) showed that the cuticular patterns on the mature leaf surfaces of *Aloë* and related genera are uniform within species, and under precise genetic control.

Leaves may also have a covering of wax over the cuticle (Fig. 4.7). The wax is either in the form of a surface crust, or more commonly in small particles of varying shapes and sizes, ranging from flakes to filaments and granules. Wax particles are variously orientated and may be in characteristic patterns. Barthlott and Wollenweber (1981) showed that many aspects of the morphology and chemistry of epicuticular waxes have taxonomic applications, particularly the orientation and distribution patterns of the particles. To some extent certain compounds, such as terpenes and flavonoids, can be recognised by their fine structure using the scanning electron microscope (Juniper and Jeffree, 1983).

4.3 Mesophyll

Chlorophyll is contained in chloroplasts in the mesophyll, which is the primary photosynthetic tissue of the leaf. The mesophyll is often divided into palisade and spongy tissues, although it may sometimes be relatively undifferentiated and homogeneous throughout the leaf. Cells of palisade mesophyll (usually adaxial), are anticlinally elongated and with few intercellular air spaces, whereas spongy mesophyll, which is most commonly abaxial, consists of variously shaped cells with many air spaces between them. Both palisade and spongy tissues may be from one to several cell layers thick, and there is sometimes intergradation between the two tissues. Many tropical grasses, and also other unrelated taxa (both monocotyledons and dicotyledons) have a ring of mesophyll cells radiating from the vascular bundles. This structure is commonly associated with the C4 pathway of photosynthesis (section 4.4.1). In thick leaves, particularly those of some monocotyledons, the central cells are often large, undifferentiated and non-photosynthetic. In the thick "keel" or "midrib" of *Crocus* leaves there is a region of large colourless cells with their walls often broken down to form a cavity (Fig. 4.8). This causes the characteristic white stripe in the "midrib" region of *Crocus* leaves.

Some xeromorphic plants, such as species of *Ilex* and many other genera, have a hypodermis, which occurs immediately within the adaxial (or more rarely the abaxial) epidermis. This consists of one or more layers of non-photosynthetic cells which are usually slightly larger

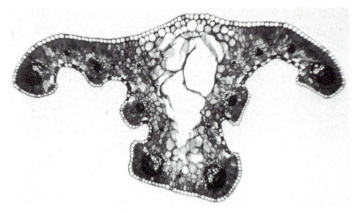

Fig. 4.8 *Crocus cancellatus*. Cross section of leaf. (x 40)

and thicker walled than the adjacent mesophyll cells, and often similar in transverse section to the epidermal cells, although without the specialised elements of the epidermis.

4.3.1 Sclerenchyma and idioblasts

Mesophyll may be interspersed with areas of sclerenchyma, particularly at the margins and extending as girders from the vascular bundles to the epidermis. Sclerenchyma, a strengthening tissue, consists of thick-walled lignified cells which generally fall into one of two categories: fibres, which are elongated and usually pointed at both ends, or sclereids, which are isodiametric or variously shaped (section 1.5). Fibres are typically found in groups associated with the vascular bundles or margins, but sclereids are often isolated in the mesophyll, and may be classified into different types depending on their shape. For example, sclereids which are more or less star-shaped with several arms (found in leaves of many taxa, e.g. *Nymphaea*: Fig. 1.6) are astro-sclereids, and those which are bone-shaped with a central column and branched ends (Fig. 4.9) are termed osteosclereids. However, many sclereids do not fit readily into any categories due to their often bizarre shapes. In some cases (e.g. in leaves of *Memecylon* and *Mouriri*) sclereids are associated with veinlet endings. Many investigations have been carried out on sclereids in leaves of vascular plants, often with taxonomic applications (e.g. Foster, 1947, 1956. See Metcalfe and Chalk, 1979, for a survey of records of sclereids in dicotyledons). Highly

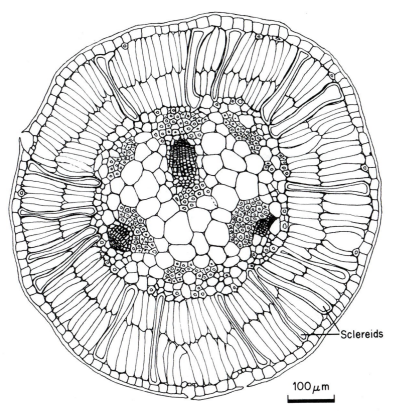

Fig. 4.9 *Hakea* sp. Leaf, cross section.

branched sclereids grow intrusively into the intercellular air spaces between neighbouring cells in the developing leaf.

Other types of idioblast may also be interspersed in the mesophyll. For example, secretory myrosin cells are often found in the leaves of Cruciferae. Laticifers, which are individual cells (non-articulated laticifers) or networks of cells (articulated laticifers) that secrete latex, occur in leaves of *Euphorbia*, *Allium* and many other genera (Fahn, 1979). Other plants may have individual specialised mesophyll cells (idioblasts) containing substances such as tannin, oil, mucilage or crystals.

Fig. 4.10 *Hyptis vauthieri.* Cleared leaf with reticulate venation.

4.4 Vasculature

The term venation is usually applied to the arrangement of the vascular bundles (veins). There are two main venation types among the angiosperms: parallel and reticulate, the former being broadly typical of monocotyledons and the latter of dicotyledons, although there are many exceptions to this.

In leaves with parallel venation the main veins are parallel for most of their length and merge or fuse at the leaf tip. There are typically numerous small veins interconnecting the larger veins, but very few vein endings in the mesophyll. In leaves with reticulate venation (Fig. 4.10) there is often a major vein in the middle of the leaf, the midrib or primary vein, which is continuous with the major venation of the

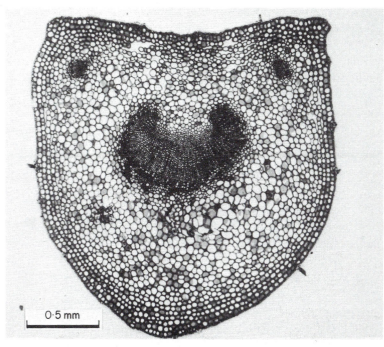

0·5 mm

Fig. 4.11 *Prunus lusitanica* (laurel). Cross section of petiole.

petiole. The midrib has smaller secondary (or second order) veins branching from it, often extending to the margins, and sometimes terminating in a hydathode. Smaller veins may in their turn branch from the second and subsequent order venation, forming a network. The areas of mesophyll between the smallest veins in the leaf are termed areoles, and in many cases small veins branch into the areoles to form vein endings. There are many variable aspects of venation: the relative number of veinlet endings per areole, or whether second order veins terminate at the margins or loop around to link with the superadjacent secondary veins. The patterns of venation in dicotyledons have been variously classified (e.g. by Hickey, 1973, 1979) and are often characteristic of taxa.

Among reticulate-veined dicotyledons the simplest form of petiole vasculature appears in cross section as a crescent, with xylem on the adaxial side and phloem on the abaxial side (Fig. 4.11). In some cases there may be additional bundles outside the main crescent, which may itself be inrolled at the ends, or in a ring, or divided into separate bundles. The classification of these various forms depends on how petiole vasculature is linked with stem vasculature at the node (Howard, 1974, 1979). One or more traces may depart from each gap in the stem vascular cylinder (section 2.2.1). The number and pattern of vascular bundles sometimes vary along the length of the petiole, and midrib vasculature, which is continuous with that of the petiole, is subject to similar variation.

In cross sections of the lamina, vascular bundles are usually in a single row, although very thick leaves, such as in some species of *Agave*, may have two or more rows. Lamina bundles are usually collateral, with adaxial xylem and abaxial phloem, but orientation may vary, and in some cases bundles may be bicollateral or even amphivasal. In the isobilateral leaves of some monocotyledons (Fig. 4.2) there are two rows of vascular bundles with opposite orientation to each other, the xylem poles towards the leaf centre. Centric leaves have a ring of vascular bundles (Fig. 4.9).

The xylem conducting system of the leaf blade may consist entirely of tracheids, usually with helical or annular thickenings, although in some leaves vessel elements and xylem parenchyma are also present. The smallest vascular bundles often have only one or two rows of xylem tracheids and a few files of phloem sieve tube elements.

4.4.1 Bundle sheath

Most minor vascular bundles in angiosperm leaves are surrounded by a bundle sheath, even to the very smallest veins. The bundle sheath typically consists of thin-walled parenchymatous cells, often in a single layer and sometimes containing starch or chloroplasts. In some monocotyledons there are distinct inner and outer sheaths, the outer bundle sheath being parenchymatous and the inner sheath sclerenchymatous and often discontinuous, forming a sclerenchyma cap usually at the phloem pole. Grasses either have a single sheath consisting of an outer layer of thin-walled cells containing chloroplasts, or a double sheath consisting of an outer layer of thin-walled cells and an inner layer of thicker-walled cells. This is an important taxonomic character in Gramineae, as double sheaths often occur in Festucoid grasses and single sheaths in Panicoid grasses, although there are occasional exceptions. Grasses from warm temperate areas with the C4 pathway of photosynthesis often display the "Kranz" type of leaf anatomy, with elongated mesophyll cells radiating from the vascular bundle sheaths, which consist of a single layer of large thin-walled cells

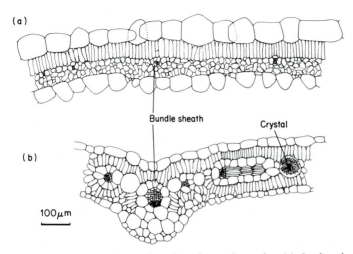

Fig. 4.12 Cross sections of leaves in region of secondary veins. (a) *Oxalis tuberosus*, (b) *Amaranthus caudatus*, leaf with Kranz anatomy.

containing chloroplasts and starch. Kranz leaf anatomy also occurs in many other plants with the C4 pathway of photosynthesis, including both dicotyledons and monocotyledons (Fig. 4.12b). Other features of C4 plants include low interveinal distance and larger chloroplasts.

Leaves of many plants have areas of sclerenchyma or parenchyma extending from the vascular bundle sheaths towards either or both epidermises. These bundle sheath extensions, also called "girders" if they reach the epidermis, afford mechanical support to the leaf, and are often regarded as a xeromorphic feature (section 4.5).

4.5 Xeromorphy

Many anatomical features, especially sunken stomata and well-developed sclerenchyma, are termed xeromorphic because they are typical of certain plants that grow in dry (xeric) environments. Similarly, other features, such as poorly-developed sclerenchyma and large air spaces in the ground tissue (aerenchyma) are often associated with water plants (hydrophytes). Xeromorphic adaptations are generally considered to be effective in reducing water loss, protecting against excessive light intensity, and in affording mechanical support. However, the physiological bases for such adaptive features in leaves are often assumed rather than proved, since some mesomorphic plants can

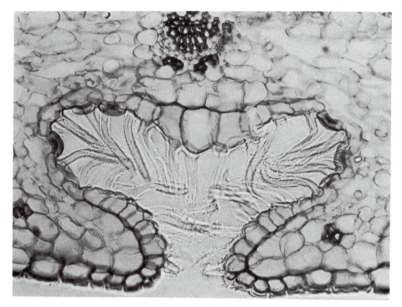

Fig. 4.13 *Erica carnea.* Cross section of leaf, showing abaxial depression lined with unicellular hairs and stomata. (x 350)

survive or even thrive in dry conditions alongside xeromorphic plants, and conversely many xeromorphic plants can occur in wet habitats. Other factors must be taken into account, such as differences in life cycle and available nutrients. For example, a mesomorphic plant may survive a period of drought in the form of a seed, and a xeromorphic plant that grows in a wet habitat may be subject to some other nutrient deficiency.

Xeromorphic leaves are often very thick, with reduced surface/ volume ratio, although some xeromorphic plants have large leathery leaves. In some cases stomata are restricted to the abaxial surface, often sunken or in grooves or depressions and surrounded by hairs (Figs 4.5b, 4.13), with the effect of creating a pocket of water vapour, and thus reducing water loss by transpiration. Other xeromorphic characters include the presence of a hypodermis or thick epidermis or cuticle, which diminish the intensity of light reaching photosynthetic tissues, and the presence of large amounts of sclerenchyma to provide mechanical support and minimise tissue collapse. Well-developed palisade tissue is often also correlated with high light intensity. Succulent plants have large thin-walled cells for water storage (see also section 2.2).

5

The Flower

5.1 Development

Studies in floral organogenesis have often revealed considerable variation between taxa (Sattler, 1973). For example, in syncarpous gynoecia the carpels may be congenitally fused so that the gynoecium arises as a single structure, or they may be initiated separately and become fused during development. Gynoecia of many species of angiosperms may therefore be apocarpous at initiation, then later become syncarpous for all or part of their length, and finally the carpels may again fall apart after fertilisation and prior to dispersal (Endress *et al.*, 1983).

Another aspect of floral variation that requires an ontogenetic approach is the sequence of initiation of the floral organs. In most angiosperm flowers the organ primordia arise from the outside inwards, in centripetal (acropetal) order. However, there are a few taxa in which all or some of the organs are initiated in the reverse order, centrifugally, and where this occurs, such as in the stamens of some groups of palms (Uhl and Moore, 1977), it may be highly characteristic. It is often impossible to determine from mature structure whether stamen initiation is centripetal or centrifugal. In *Drimys* the innermost stamens are the largest and the first to dehisce, but their initiation is centripetal (Tucker, 1972).

5.2 Vasculature

Floral vasculature has been studied extensively in the past, often based on the (now controversial) theory of vascular conservatism, which held

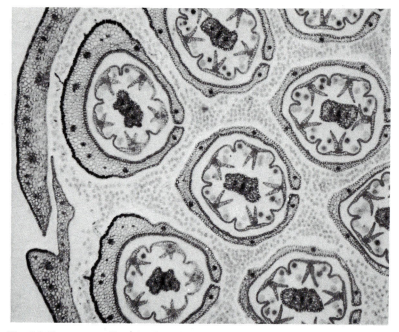

Fig. 5.1 *Taraxacum officinale* (dandelion). Cross section of inflorescence, showing individual florets surrounded by outer petals. (x 40)

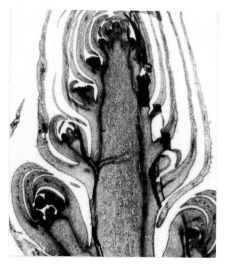

Fig. 5.2 *Caesalpinia yucatanensis*. Longitudinal section of young inflorescence, showing floral primordia. (x 50)

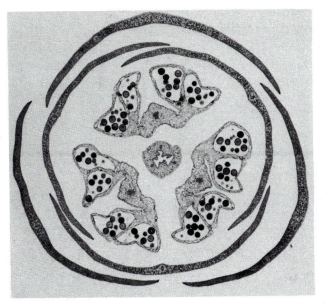

Fig. 5.3 *Crocus sieberi.* Cross section of flower, with six outer tepals surrounding three anthers and central style. (x 40)

that the presence of vestigial vascular bundles in certain parts of the flower can be taken to represent vestigial organs, and hence elucidate problems of floral evolution (Puri, 1951). However, as Schmid (1972) pointed out, the degree of fusion of floral vasculature can sometimes vary even within a single species.

In most flowers the vascular traces to each organ diverge from the central vascular cylinder in a series of whorls, at different levels in the flower. The number of traces to petals and sepals is very varied in different flower types, and the perianth traces often branch dichotomously, as in the leaf. In stamens the most common type of vasculature is a single vascular strand (Figs 5.3, 5.5), but some families characteristically have three or four stamen traces, and others, such as Araceae (French, 1986), have diverse and often branching stamen vasculature. The carpellary vascular system is divided into ventral carpellary traces, which diverge into the ovules, and dorsal carpellary traces, which pass up the style into the stigma (Fig. 5.9). The number of vascular bundles in the style of a syncarpous gynoecium is often an indicator of the number of carpels, although sometimes bundles may be branched or fused.

In flowers with very complex vascular systems, such as those of many Magnoliaceae, vasculature is most easily studied in the immature bud, where the ontogenetic sequence of initiation can be observed. Some authors have suggested that xylem and phloem initiation may be related to functional aspects; for example, phloem may develop precociously in areas which require a copious supply of nutrients, such as developing sporogenous tissue.

5.3 Petals and sepals

In their simplest form, perianth parts are essentially leaf-like, although there are many flowers with modified and fused petals, and sometimes fused sepals. In cross section, both petals and sepals consist of an abaxial and adaxial epidermis enclosing usually three or four or sometimes up to ten layers of undifferentiated isodiametric or elongated cells with many air spaces between them. This "mesophyll" tissue is interspersed with a row of vascular bundles. In general, petals and sepals show fewer modifications than leaves in their internal structure, although they may also contain idioblasts such as crystal-containing cells, or specialised tissues such as a hypodermis. Sepals are frequently green and photosynthetic. It is common to find stomata and trichomes on sepal surfaces, and also sometimes on petals.

In insect-pollinated flowers the main function of the corolla is usually to attract insects, and it is consequently the largest and showiest part of the flower. (In wind-pollinated flowers the reverse is often the case, the perianth being often much reduced or even absent.) Flower colour is mainly controlled by the chemistry of the pigments. Kay *et al.* (1981), in a survey of 60 plant families, showed that with few exceptions anthocyanins, betalains and ultraviolet-absorbing flavonoids are confined to the epidermal cells of the petal, whereas other pigments, such as carotenoids, are found in either the epidermis or the mesophyll. They correlated this information with observations on the surface morphology of the petals. Petal surfaces are either smooth (as in species of *Crocus*), in which case incident light is reflected strongly, or papillate (as in species of *Viola*, *Veronica* or *Cistus*) with papillae of various heights, and either one or several per cell (Fig. 5.4). The effect of the domed or papillate cell surface is to guide incident light into the petal, where it is reflected outwards from the inside walls of the epidermal cells or from the multi-faceted walls of mesophyll cells, thus passing through the

Fig. 5.4 *Veronica* sp. Petal surface, showing papillate cells with striations. (SEM, x 210)

pigments in solution in the cell vacuoles. In some species of *Ranunculus* with bright yellow flowers, incident light is reflected from starch grains in the subepidermal mesophyll cells.

Many petal surfaces are also strongly striated, which may have the effect of further scattering the incident light into the interior of the petal. Striations on papillae usually run from the base to the apex of the papilla. Gale and Owens (1983) found that in some genera of Commelinaceae petal surfaces were smooth, and in others striated. They also found that cuticular striations occurred on the stamens, staminodes and styles of many species, indicating that these organs also play a part in light absorption patterns and attraction of insects to the flower.

5.4 Androecium

5.4.1 Stamens

Stamen filaments are usually slender and cylindrical, but may be leaf-like or branched. In cross section they have a parenchymatous ground tissue surrounding the usually simple vasculature (section 5.2). The

Tapetum Pollen Anther locule   Vascular bundle Epidermis Endothecium

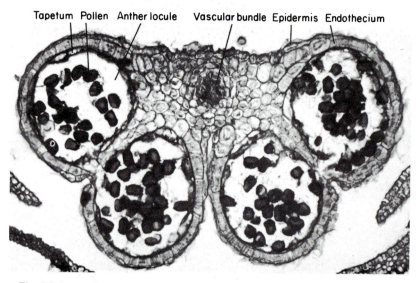

Fig. 5.5 *Sisyrinchium* sp. Cross section of anther. (x 150)

filament surface often bears trichomes, stomata and surface patterning, as in other floral parts.

The anther wall is made up of several layers which are all derived from primary parietal (wall) cells, apart from the epidermis, which undergoes only anticlinal divisions during development. The two most distinct layers are the endothecium (Fig. 5.5), immediately within the epidermis, and the tapetum, which is the innermost layer of cells surrounding the anther locule. Intervening layers usually consist of thin-walled cells which are often crushed and destroyed at anthesis, although Davis (1966) recorded various types of inner parietal layers. Endothecial cells, most commonly in a single layer, typically develop fibrous wall thickenings which contribute to the anther dehiscence mechanism. Tapetal cells are secretory and full of dense cytoplasm. The contents of the tapetal cells are absorbed by the developing pollen grains in the anther locule, so that by the time the pollen grains are mature the tapetum has usually completely degenerated. In many cases the tapetal cells break away into the anther locule, either as amoeboid cells, or sometimes their protoplasts coalesce to form a periplasmodium, which also eventually disintegrates.

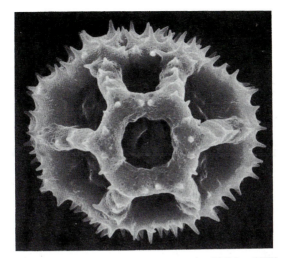

Fig. 5.6 *Taraxacum officinale* (dandelion). Pollen grain. (SEM, x 2000)

5.4.2 Pollen

In the developing anther the primary sporogenous cells, which are derived from the same initials as the primary parietal cells, give rise either directly or by successive mitotic divisions to the microspore parent cells. These in turn each undergo two meiotic divisions (microsporogenesis), either successive or simultaneous, to form a tetrad of haploid microspores. Individual microspores usually separate at this stage, although in some families they remain together as permanent tetrads. Prior to release of pollen from the mature anther, the microspores undergo an unequal division to form a larger vegetative cell and a smaller generative cell enclosed within the pollen grain wall. The generative nucleus later further divides to form two sperm nuclei, but this division usually occurs in the pollen tube. The mature pollen grain, the male gametophyte, is therefore usually bicellular (or rarely tricellular).

Pollen grains (Fig. 5.6) are radially or bilaterally symmetrical bodies that represent highly adapted units of dispersal from the anther to the stigma. The pollen grain wall has two main layers, the soft inner intine and the hard outer exine, the exine being further subdivided into the outer sculptured sexine and the inner non-sculptured part, the nexine. In most cases the exine has openings, or apertures, although there are rare examples of non-aperturate (inaperturate) pollen grains. Apertures may be elongated furrows or more or less circular pores, or may sometimes have a more complex inner and outer arrangement (Huynh, 1976). In sulcate pollen grains, found in many monocotyledons, the apertures (sulci) lie along the distal face (the face that was directed outwards in the tetrad). In colpate pollen, the furrows (colpi or pores) are equatorial, as defined during the tetrad phase. Spiraperturate pollen grains have spiral apertures.

Pollen grains vary considerably in size and shape, and there are many different patterns of sexine sculpturing. For example, the sexine may be reticulate or areolate, or with surface holes (puncta), granules, warts or spines (Erdtman, 1966). These differences are often of considerable taxonomic significance either at or below the family level. In many taxa this surface patterning may be a mechanical adaptation, ensuring elasticity of the wall, or providing resistance to collapse. Pollen grains dehydrate after contact with the air, and the exine contracts. On the stigmatic surface rehydration and exine expansion occurs. Dry and hydrated pollen grains of the same species can appear very different in size, shape and even surface features. Wind-borne pollen grains are generally small and light, and with low surface sculpturing, and water-dispersed pollen may have adaptations such as a slime coating (e.g. in Hydrocharitaceae: Knox, 1984). In some species, especially those with animal-dispersed pollen, various substances such as lipids, proteins and carbohydrates are stored and dispersed with the pollen in the inter-columellar spaces of a deeply chambered exine. The substances are often derived from the tapetum, and may have various functions, such as conferring odour, or causing grains to aggregate into sticky masses, useful for effective animal dispersal. They may also have a role in the control of interspecific compatibility, especially in certain families, such as Cruciferae, Malvaceae or Compositae (Heslop-Harrison, 1976), as they are released on the stigma when the exine expands after pollen grain rehydration.

5.5 Gynoecium

5.5.1 Stigma and style

Stigmatic epidermal cells are modified to provide a receptive surface for pollen grains. They are secretory, with a specialised cuticle, and may be slightly domed or with variously elongated papillae. Stigmas can broadly be divided into two types: "dry", with little or no surface

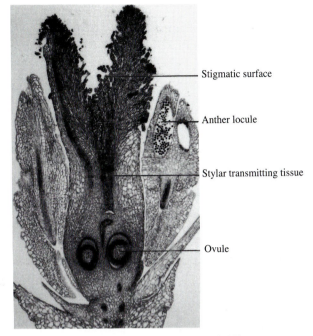

Stigmatic surface

Anther locule

Stylar transmitting tissue

Ovule

Fig. 5.7 *Flagellaria indica.* Longitudinal section of open flower. (x 30)

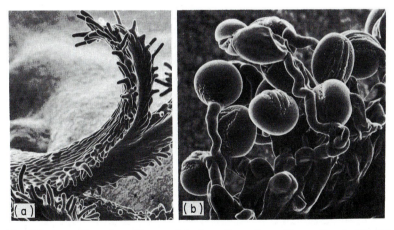

Fig. 5.8 *Anomatheca laxa.* Flower: (a) stigmatic surface with papillae (x 53), (b) germinating pollen grains on stigma (x 230).

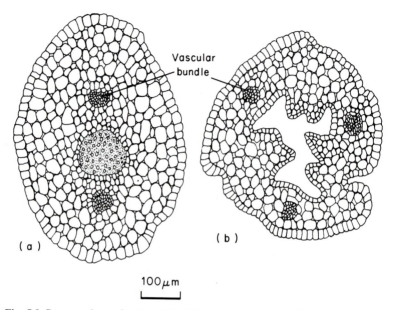

Vascular bundle

100 μm

Fig. 5.9 Cross sections of styles of (a) *Salvia pratensis* ("closed" style), (b) *Crocus speciosus* ("open" style).

secretions, or "wet", with surface secretions which assist in pollen germination (Heslop-Harrison and Shivanna, 1977). The stigmatic cuticle usually appears stratified in cross section, with a lamellated outer layer and reticulate inner layers; and there are many modified types, such as in *Crocus*, which has a chambered cuticle, and *Euphorbia*, where the cuticle is fenestrated (Heslop-Harrison and Heslop-Harrison, 1982). The cuticle is often torn at maturity as secretions are released. The germinating pollen tubes (Fig. 5.8) grow down either between the stigmatic papillae or along the surface of the stylar canal.

The anatomy of the style varies depending on the degree of fusion of the carpels. In many of the higher angiosperms with syncarpous gynoecia the style is solid, with a central specialised secretory tissue, the transmitting tissue, which links the stigma with the centre of the ovary, and serves as a nutrient-rich tract for pollen tube growth (Fig. 5.9a). Other flower types have a central stylar canal lined with secretory, often papillate epidermal cells (Fig 5.9b), and in a few cases the stigma is borne directly on the top of the ovary. In a comprehensive study of the gynoecium of *Ornithogalum caudatum*, Tilton and Horner (1980) observed that the tissue traversed by the pollen tubes from the top of the

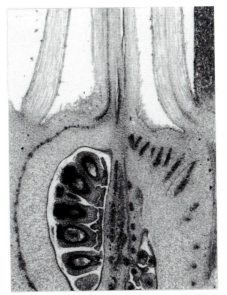

Fig. 5.10 *Bomarea hirtella.* Longitudinal section of ovary and lower part of style. (x 15)

stigma to the base of the locule in the ovary is morphologically uniform, and derived from the adaxial epidermis of the carpel or its immediate derivatives.

The ground tissue of the style is parenchymatous, interspersed with a number of vascular bundles (section 5.2).

5.5.2 Ovary

The ovary contains one or more ovules (Fig. 5.7, 5.10, 5.11), and each ovule is attached to the ovary wall at a placenta, of which there are usually two per carpel. The term "placentation" refers to the arrangement of the placentae and the number of carpels in the ovary, both of which depend on the degree of fusion of the carpels. In syncarpous ovaries, where the carpels are congenitally fused or have grown together postgenitally so that they are fused in the centre of the ovary, there are usually as many locules as carpels, and placentation is axile. Where the carpels are fused at their outer edges but not in the centre of the ovary, there is usually only one locule and placentation is parietal (although rarely false septa occur). Other placentation types may occur in

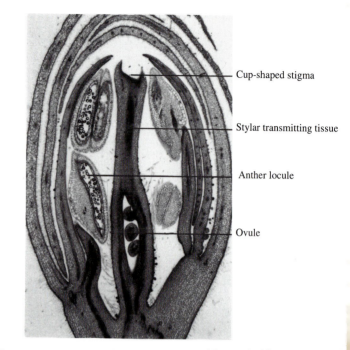

Cup-shaped stigma

Stylar transmitting tissue

Anther locule

Ovule

Fig. 5.11 *Caesalpinia mexicana*. Longitudinal section of flower. (x 30)

unilocular ovaries; for example, the placentae may be at the base of the ovary (basal placentation), or on top of a central column of tissue that is not joined to the ovary wall except at top and bottom (free central placentation).

In species with solid styles, transmitting tissue is continuous from the style through the centre of the ovary, for the growth of pollen tubes to the ovules. There is often an opening (compitum) near the micropyle of each ovule, which enables pollen tubes to grow into an embryo sac. Many taxa have obturators, which are proliferations of secretory tissue around the bases of the funicles, providing nutrients for the developing pollen tubes and guiding them into the micropyles.

5.5.3 Ovule

Each ovule is attached to the ovary by a funicle (or funiculus), which has a single vascular strand. The ovule wall, which will develop into the seed coat after fertilisation, consists of an inner layer, the nucellus,

Micropyle

Nucellus

Megaspore mother cell

Inner integument

Outer integument

Fig. 5.12 *Chondropetalum tectorum*. Cleared developing ovule at megaspore mother cell stage, with growing integuments; micropyle not yet fully formed.

surrounded by one or two (inner and outer) integuments (Figs 5.12, 5.13). The region where the nucellus, integuments and funicle unite is termed the chalaza (Fig. 6.1), usually opposite the micropyle, which is a narrow opening at one end of the ovule, formed by one or both integuments. During the process of fertilisation, pollen tubes pass through the micropyle into the embryo sac. Like the tapetum in the anther, the nucellus is often absent at maturity, having degenerated, but at the onset of megasporogenesis it consists of either a single layer of cells, or several layers with a well-defined inner and outer epidermis. In some mature ovules the nucellus may have proliferated at the chalazal end of the ovule, opposite the micropyle, to form a hypostase. The hypostase is often refractive and suberinised or lignified, and it may be very large, as in *Crocus* (Fig. 5.13b). The vascular bundle in the funicle usually terminates at the chalaza, but in some cases it is more extensive.

5.5.4 Embryo sac

Maheshwari (1950), Davis (1966) and Willemse and Van Went (1984) summarised the many variations in embryo sac development in angiosperms. As in microsporogenesis, the primary sporogenous or archesporial cells give rise either directly or by successive mitotic divisions to the megaspore mother cell. This then undergoes mega-

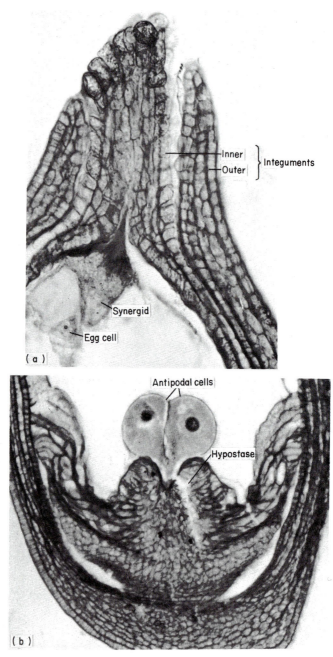

Fig. 5.13 *Crocus sieberi.* Longitudinal sections of ovule at (a) micropylar end (x 340), (b) chalazal end (x 135).

sporogenesis, with two meiotic divisions to form a tetrad of four megaspores of varied arrangement and shape (usually linear or T-shaped). In the majority of angiosperm flowers, one megaspore (often the chalazal one) then gives rise to the mature embryo sac by further mitotic divisions, and the other three megaspores degenerate, although there are many instances where more than one megaspore has a role in embryo sac formation.

In most angiosperms the mature embryo sac (megagametophyte) has eight nuclei, although types with four and sixteen or more nuclei have also been recorded. These eight nuclei consist of three antipodals (each usually with a cell wall) at the chalazal end, two polar nuclei towards the centre, and two synergids plus an egg cell (the egg apparatus) at the micropylar end. Many mature embryo sacs have only seven nuclei because the two polar nuclei fuse at an early stage to form a diploid fusion nucleus, which will eventually fuse with a male sperm nucleus to form the triploid endosperm. The synergids often have wall thickenings, the filiform apparatus, which extend into the micropyle (Fig. 5.13a).

5.6 Floral secretory structures

Flowers often bear specialised secretory structures, such as nectaries, elaiophores and osmophores, mainly to attract potential pollinators, especially insects such as bees and moths, but also humming birds and bats. Nectaries are clearly localised areas of tissue that regularly secrete nectar, a sugary substance which is attractive to animals (Daumann, 1970). Schmid (1988) summarised the complex terminology relating to both floral and extrafloral nectaries and related secretory structures. Floral nectaries may occur on the petals, or on any other floral parts. For example, in many Labiatae and other dicotyledons an enlarged "disc" at the base of the ovary forms a nectary. Nectaries are sometimes found in special nectar-collecting furrows such as the septal nectaries of some monocotyledons, in the slits that result from the (normally fused) carpel margins being separate at the edge of the ovary. Nectaries usually consist of secretory epidermal cells with dense cytoplasm, sometimes modified into trichomes. The subepidermal cells may also be secretory, and in some cases nectar passes to the surface through modified stomatal pores. Vascular tissue close to the nectary often consists mainly or entirely of phloem, which transports sugars to the secretory region.

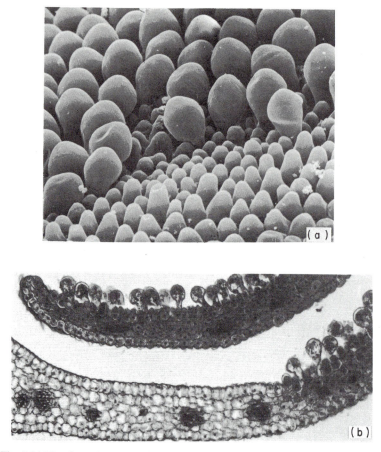

Fig. 5.14 *Tigridia meleagris.* Elaiophores on petal surface. (a) SEM, x 300, (b) cross section, x 130.

Some flowers bear oil-secreting glands, termed elaiophores (Fig. 5.14), which are also attractive to insects, and morphologically similar to some types of nectary (Vogel, 1974). Osmophores, which produce volatile secretions (scents) that are attractive to potential pollinators, consist of a diffuse region of secretory epidermal and underlying tissue (Vogel, 1961).

6

Seeds and fruits

6.1 Seed coat

Most angiosperm seeds have a seed coat, or testa, derived from the two integuments or single integument of the ovule wall (Figs 6.1, 6.2). In bitegmic seeds the term "testa" is correctly applied only to the outer layer, formed from the outer integument, the part formed from the inner integument being the tegmen. The nucellus often degenerates at an early stage, but may persist, especially at the chalazal end, where it sometimes forms a food storage tissue, the perisperm.

Seed coats may be complex multilayered tissues, or simply enlarged ovule walls. They generally include a hard, protective layer formed from all or part of the testa or tegmen. Corner (1976) classified seed coats according to the position of this mechanical layer. For example, in exotestal seed coats the mechanical layer is formed from the outer epidermis of the outer integument, and in endotegmic seed coats it is derived from the inner epidermis of the inner integument. Sometimes the mechanical layer consists of one or more rows of elongated, palisade-like cells, such as the macrosclereids in the exotesta of many Leguminosae.

Apart from the obvious mechanical protective function, to prevent destruction of the seed by dehydration or predation, the seed coat often has important subsidiary functions, usually related to dispersal, and with corresponding specialised structures (Boesewinkel and Bouman, 1984). For example, seeds may be winged, for wind dispersal, or fleshy, for dispersal by animals. The fleshy part of the seed coat, most commonly formed from part of the outer integument, is termed the sarcotesta, although arils are also fleshy outgrowths of the seed (from the funicle), and display a great variety of forms. The vasculature of the

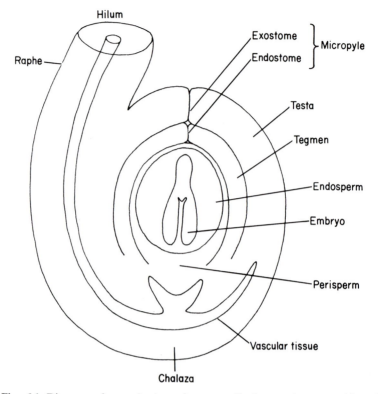

Fig. 6.1 Diagram of organisation of a generalised campylotropous bitegmic dicotyledon seed with perisperm. (After Boesewinkel and Bouman, 1984.)

seed coat usually consists of a single bundle passing from the raphe to the chalaza, but this can vary in extent and degree of branching. In orchid seed coats vasculature is often completely absent.

Seed coat surfaces exhibit a variety of cellular patterns, often with papillate or striate surface sculpturing which may be characteristic of taxa (Fig. 6.3) (Barthlott and Ehler, 1977; Barthlott, 1981). Many seeds have epidermal trichomes, notably *Gossypium* (cotton), where the epidermal outgrowths are an important source of textile fibres. In mucilage seeds and fruits, of which there are many examples among angiosperms, the epidermal cells of the seed coat or fruit wall absorb water and rupture, producing large amounts of slime interspersed with coiled thread-like protruberances (e.g. in the nutlet walls of *Coleus* and some other Labiatae). The main purpose of this mucilage production is

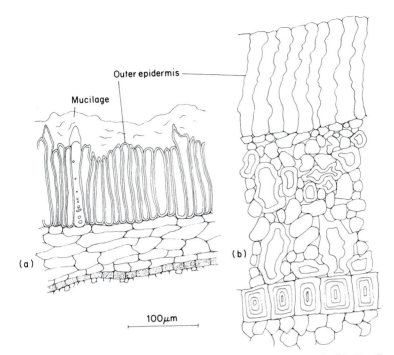

Fig. 6.2 Cross sections of testa of (a) *Citrus sinensis* (sweet orange), (b) *Citrullus vulgaris* (water melon).

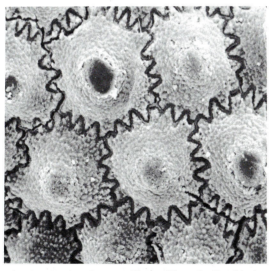

Fig. 6.3 *Silene nutans*. Seed surface; papillate epidermal cells with sinuous anticlinal walls. (SEM, x 260)

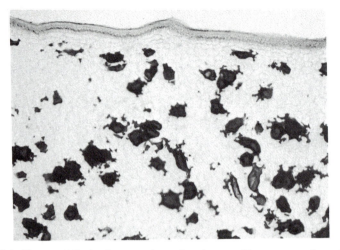

Fig. 6.4 *Olea* sp. (olive). Cross section of pericarp, with numerous branching sclereids. (x 60)

unknown, but it may be simply to adhere the seed firmly to the soil surface, or to the surface of an animal for dispersal.

6.2 Fruit wall

The fruit wall is called the pericarp if the fruit develops from a single ovary and the fruit wall from the ovary wall. It displays a similar range of variation to the seed coat, depending particularly on whether the fruit is dry or fleshy. The pericarp is typically divided into three layers: the outer exocarp, central mesocarp and inner endocarp, although sometimes the three layers are not readily distinguishable. As with the seed coat, at least one layer of the fruit wall often consists of thick-walled lignified cells, although in some fleshy fruits (berries), such as the grape, the entire endocarp may consist of thin-walled succulent cells. In other fleshy fruits (drupes), such as the peach, the endocarp cells are thick walled, and only the mesocarp is fleshy, the exocarp being a narrow epidermal layer. In the olive the fleshy mesocarp is interspersed with thick-walled sclereids (Fig. 6.4).

6.3 Endosperm

The endosperm is usually triploid, formed by fusion of one male nucleus with two female polar nuclei. It is a nutritive, food-storage tissue, present in most angiosperm seeds but in greatly varying amounts, and occasionally absent (e.g. in Orchidaceae). There are three main types of endosperm: nuclear, cellular and helobial, defined by differences in early development. In nuclear endosperm the early cell divisions are unaccompanied by cytokinesis (cell wall formation), and the nuclei are initially free in the cytoplasm of the embryo sac, usually surrounding a central vacuole. In most cases of this type, walls do form eventually, but sometimes the nuclei at the chalazal end remain free; for example coconut palms (*Cocos nucifera*) have nuclear endosperm formation, and the liquid "milk" contains many free endosperm nuclei. In the cellular type of endosperm formation, cytokinesis occurs even with the initial cell divisions, although the pattern of cell wall formation varies considerably between different taxa. Both nuclear and cellular types are widespread among angiosperms. Helobial endosperm formation occurs only in monocotyledons, and is sometimes considered an intermediate type. The division of the primary endosperm nucleus is accompanied by the formation of a small chalazal chamber and a large micropylar chamber. The nucleus of the micropylar chamber migrates to the top of the embryo sac, and its initial divisions are unaccompanied by cytokinesis, although with later mitoses cell walls are formed. The chalazal chamber has far fewer nuclear divisions, and its nuclei remain free in the cytoplasm. Vijayaraghavan and Prabhakar (1984) recognised several types of helobial endosperm formation.

In many families the endosperm develops haustoria, which help in nutrient absorption, sometimes invading surrounding tissues (Maheshwari, 1950; Vijayaraghavan and Prabhakar, 1984). For example, most species of the families Labiatae and Verbenaceae have both chalazal and micropylar haustoria (Fig. 6.5), which may be either free-nucleate or cellular, sometimes even amoeboid. In these taxa, the first division of the fusion endosperm nucleus is longitudinal, followed by the formation of a transverse wall. The chalazal nucleus forms a small chalazal haustorium close to the antipodals, and the micropylar nucleus divides further to form a micropylar haustorium and a central cellular endosperm. The micropylar haustorium transfers nutrients from the integument to the embryo and cellular endosperm. The chalazal haustorium transfers nutrients from the vascular bundle to the endosperm.

Mature endosperm typically consists of tightly packed cells with either thick or thin walls, and containing food reserve materials such as starch grains or protein bodies, which may be in various characteristic forms.

6.4 Embryo

In a normal angiosperm reproductive system the embryo develops from the diploid fertilised egg cell (zygote). Following fertilisation the zygote often undergoes a change in volume, either shrinkage or enlargement, before cell division commences, usually initially by a transverse division to form apical and basal cells. The pattern of subsequent cell division varies between taxa and has been classified into several types (Maheshwari, 1950). In many cases considerable taxonomic significance is attached to the type of early embryogenesis, and there are distinct differences between the early embryogenesis of dicotyledons and monocotyledons. However, most embryos eventually differentiate into a globular mass of cells (the proembryo) attached to the embryo sac wall by a stalk (the suspensor) (Fig. 6.5). The suspensor also exhibits great diversity, and sometimes has a secretory function. It may be uniseriate or multiseriate, but ultimately degenerates.

The undifferentiated globular proembryo may be massive (e.g. in *Degeneria*), or small, as in *Capsella*, where it consists of only eight cells (Steeves and Sussex, 1989). However, by a process of irregular meristematic activity, it eventually becomes organised into a structure with root and shoot apices at opposite ends of an axis (the hypocotyl). Many dicotyledon embryos become bilobed (Fig. 6.1) as the two cotyledons differentiate, and monocotyledon embryos develop a single, often elongated cotyledon. The degree of differentiation of mature embryos varies considerably; in orchids the embryo remains a simple undifferentiated mass of cells. Some highly differentiated embryos may have, in addition to the hypocotyl and cotyledons, a short primordial root (radicle), often with a root cap, and a shoot bud or short shoot (the epicotyl) developed beyond the cotyledons. Grass embryos (Fig. 6.6) are unique highly differentiated structures with a long specialised cotyledon (the scutellum), and sometimes an outgrowth opposite the scutellum, the epiblast, which has been variously interpreted as a second cotyledon or an outgrowth of the first cotyledon or the coleorhiza (Natesh and Rau, 1984). Grasses characteristically have a sheath (coleoptile)

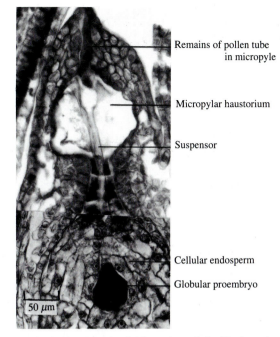

Remains of pollen tube in micropyle

Micropylar haustorium

Suspensor

Cellular endosperm

Globular proembryo

50 μm

Fig. 6.5 *Pogostemon plectranthoides.* Longitudinal section of fertilised ovule at micropylar end. (Composite picture, of two serial sections.)

surrounding the epicotyl and plumule, and a well-developed radicle also surrounded by a sheath (the coleorhiza), although Tillich (1977) found that the coleorhiza is not restricted to grasses, but is widespread among monocotyledon seedlings, occurring on every shoot-borne (adventitious) root and developing from outer cortical tissue by cell elongation.

6.5 Seedling

At germination the testa is ruptured and the radicle emerges through the micropyle. The seedling is the most juvenile stage of the plant, immediately after germination. Seedlings are fairly diverse in structure, depending on the relative size and position of the component parts. Variation in seedling morphology is often of taxonomic significance. Seedlings have a root (radicle) and a hypocotyl, which bears the cotyledons and plumular bud, although in monocotyledons the radicle withers at an early stage, and subsequent roots are shoot-borne

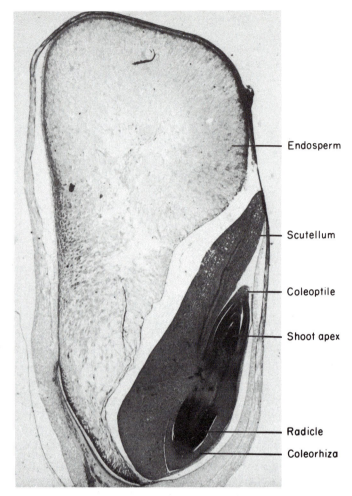

— Endosperm

— Scutellum

— Coleoptile

— Shoot apex

— Radicle
— Coleorhiza

Fig. 6.6 *Zea mays* (maize). Longitudinal section of seed. (x 12)

(adventitious). The plumular bud produces the stem and leaves, which soon resemble those of the mature plant. The hypocotyl varies in size and form, from a swollen food-storage organ to a very short structure which may be almost non-existent, in which case the radicle extends almost to the cotyledonary node. The cotyledons, or seed leaves, usually differ from the first foliage leaves; for example they often have simpler vasculature which may consist of a single vascular bundle (Fig. 6.7). In some cases, mainly in larger-seeded dicotyledons such as the legume pea

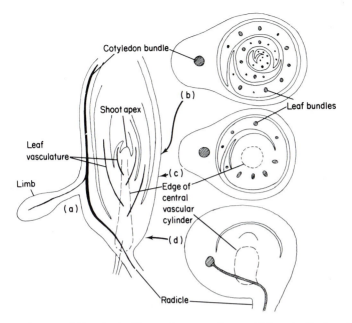

Fig. 6.7 Diagram of *Tigridia* seedling to show vasculature. (a) Longitudinal section, (b)(c) and (d) cross sections at positions indicated by arrows.

or bean (e.g. *Vicia faba*), the cotyledons are fleshy and swollen, with a food-storage function, and remain permanently enclosed in the testa even after the radicle has emerged. This is termed hypogeal germination, as opposed to the more common epigeal germination, where the cotyledons are borne above ground, and are usually photosynthetic. In monocotyledon seedlings the cotyledon typically consists of three parts: a basal sheath, a ligule or ligular sheath, and a limb (Arber, 1925), each part being more or less well developed. For example, in *Tigridia* seedlings (Fig. 6.7) both the hypocotyl and the basal sheath are extremely reduced and almost non-existent. Cotyledon vasculature is usually much simpler than that of foliage leaves, and may consist of a single vascular bundle.

Glossary

abaxial: away from the axis. (The abaxial leaf epidermis is most commonly the lower one.)

abscission layer: region of separation of tissues, e.g. for abscission of leaf from stem.

acropetal: towards the apex.

adaxial: towards the axis. (The adaxial leaf surface is most commonly the upper one.)

adventitious roots: stem or leaf-borne roots.

aerenchyma: specialised parenchymatous tissue normally associated with aquatic plants, with a regular, well-developed system of intercellular air spaces.

amphistomatic leaves: with stomata found on both surfaces.

amphivasal vascular bundle: one with xylem surrounding phloem.

amyloplast: plastid containing starch.

androecium: whorl(s) of stamens (pollen-bearing organs) in flower.

anisocytic: one of the classification types of mature stomata; with three unequal subsidiary cells.

anomalous secondary growth: secondary growth that does not fit the "normal" pattern of xylem and phloem production; e.g. xylem with included phloem.

anomocytic: one of the classification types of mature stomata; with no subsidiary cells.

anther: part of stamen bearing pollen (in anther locules).

anticlinal: perpendicular to the surface.

antipodal cell: one of a group of (most commonly three) cells at the chalazal end of the mature megagametophyte (embryo sac).

apocarpous gynoecium: one where carpels are not fused.

apotracheal parenchyma: (in secondary xylem) not associated with the vessels.

areole: region of mesophyll between smallest veins in leaf.

aril: (in seeds) fleshy outgrowth of the funicle.

articulated laticifer: one composed of several cells, their adjoining walls often broken down.

astrosclereid: star-shaped or highly branched sclereid (Fig. 1.6).

bark: the part of the woody stem outside the secondary xylem; i.e. including vascular cambium, phloem, cortex and periderm, although sometimes applied only to periderm and outer cortex.

basipetal: towards the base; i.e. away from the apex.

bast fibre: extraxylary fibre in stem; i.e. cortical or phloem fibre.

bicollateral vascular bundle: one with phloem on both sides of the xylem.

bifacial leaf: one with both adaxial and abaxial surfaces, usually differing from each other morphologically.

bitegmic seed: one with two seed coat layers, derived from two integuments.

brachysclereid: more or less isodiametric sclereid (stone cell).

bulliform cells: (usually in certain grass leaves) groups of epidermal cells that are markedly larger than neighbouring epidermal cells.

bundle sheath: sheath of cells surrounding leaf vascular bundles; may be complete or incomplete, parenchymatous or sclerenchymatous (or both), and single or double (consisting of two distinct cell types).

callus tissue: undifferentiated mass of thin-walled cells; usually wound tissue.

cambium: meristematic band of cells; e.g. cork cambium or vascular cambium.

campylotropous ovule: one which has become bent over through 90 degrees during development, and fused to the funicle.

casparian strip, or thickening: band of suberin or lignin in primary walls of cells of root endodermis.

cellulose: a carbohydrate; a crystalline compound which is the main component of plant cell walls.

centric (or terete) leaf: one that is cylindrical.

centrifugal: outwards (from the inside).

centripetal: inwards (from the outside).

chalaza: the end of the ovule or seed opposite the micropyle (Fig. 6.1); the region where nucellus and integuments merge.

chlorenchyma: photosynthetic tissue; specialised parenchyma containing chloroplasts.

chlorophyll: green photosynthetic pigment contained within chloroplasts.

chloroplast: plastid containing chlorophyll, the site of photosynthesis.

coenocyte: multinucleate cell; i.e. one where cell division has occurred without cell wall formation, e.g. non-articulated laticifer.

coleoptile: (in grass seedlings) parenchymatous sheath enclosing plumule.

coleorhiza: (in monocotyledon seedlings, especially grasses) parenchymatous sheath covering primary root.

collateral vascular bundle: one with xylem and phloem adjacent to each other.

collenchyma: strengthening tissue, consisting of groups of axially elongated, tightly packed cells with unevenly thickened walls. Usually found immediately within the epidermis at the angles or ridges of stems or petioles.

colpus: aperture in pollen grain wall, aligned equatorially as defined during the tetrad stage, and usually elongated.

companion cell: parenchymatous cell associated with sieve tube element in phloem.

compitum: opening in transmitting tissue of ovary, near micropyle.

cork: suberinised tissue (periderm).

cortex: region in stems and roots between epidermis and central vascular region.

cotyledon: first leaf of the embryo.

cuticle: non-cellular layer of a fatty substance (cutin), covering surface of epidermis.

cystolith: calcareous body found in epidermal cell, or in leaf mesophyll.

cytokinesis: cell wall formation.

diacytic: one of the classification types of mature stomata; with one or more pairs of subsidiary cells with their common walls at right angles to the guard cells.

diarch root: one with two protoxylem poles.

dictyosome (Golgi body): cell organelle, associated with secretory activity.

distal: situated away from the centre of a body or its point of attachment; terminal; (opposite of proximal).

domatia: specialised pockets or tufts of hairs on some leaf surfaces, providing shelter for small insects; sometimes associated with extrafloral nectaries.

dorsiventral leaf: one with the two surfaces morphologically different from each other.

druse: cluster crystal, or compound crystal.

ectomycorrhizal: (fungal mycelium on roots) associated with the surface.

egg apparatus: (in mature embryo sac) egg cell and two synergids.

egg cell: (in mature embryo sac) haploid cell at micropylar end that will fuse with male haploid nucleus to form a zygote, and subsequently an embryo.

elaiophore: oil-secreting trichome or tissue in flower.

embryogenesis: embryo development. Somatic embryogenesis is the induction of an embryo-like structure in cell suspension cultures and on the surface of callus cultures.

endocarp: inner layer of fruit wall (pericarp).

endodermis: (mainly in roots) innermost cell layer of cortex; initially often with Casparian strip or thickening, later often thick walled. Distinct endodermoid layer also sometimes present in stems, but usually lacking Casparian thickenings.

endogenous: of deep-seated (internal) origin.

endomycorrhizal: (fungal mycelium on roots) invading tissues and cells.

endoplasmic reticulum (ER): series of membranes permeating the cytoplasm.

endosperm: seed storage tissue, formed by fusion of one male nucleus with two female polar nuclei (i.e. usually triploid). Three main types: nuclear, cellular and helobial.

endotegmic seed coat: one with thickened, mechanical layer derived from the inner epidermis of the inner integument.

endothecium: anther wall layer immediately within the epidermis; often with characteristic thickenings (Fig. 5.5).

epiblast: in grass embryos, outgrowth opposite scutellum.

epicotyl: seedling axis above cotyledons.

epidermis: outermost layer of cells, over entire primary plant surface. Normally uniseriate, but rarely multiseriate.

epigeal germination: type where cotyledons are green and borne above ground.

epigynous flower: one with inferior ovary (i.e. the ovary is below the level of insertion of the stamens and perianth parts).

epistomatic leaves: with stomata found on adaxial surface only.

epithem: tissue (often loosely packed parenchyma) in hydathode between epidermis and vascular tissue.

exarch: maturing centripetally.

exine: outer coat of pollen grain.

exocarp: outer layer of pericarp (fruit wall).

exodermis: outer few cell layers of root cortex that have become thicker walled and lignified.

exogenous: of superficial (external) origin.

exotestal seed coat: one with mechanical layer formed from the outer epidermis of the outer integument.

fibre: axially elongated, thick-walled cell, usually occurring as part of a group, lacking contents at maturity, and with simple pits (except for xylem fibre-tracheids, a transitional type, with bordered pits). Septate fibres occur in some woods.

filament: (in flower) stalk of stamen.

filiform apparatus: wall thickenings in synergid cells of mature embryo sac.

funicle or funiculus: stalk attaching ovule to ovary.

fusiform: elongated with pointed ends.

generative cell: part of male gametophyte; divides (usually within pollen tube) to form two sperm cells.

girder (or bundle sheath extension): in leaves, usually a group of cells (parenchymatous or sclerenchymatous) linking a vascular bundle sheath with either or both epidermises.

graft: union (by cell differentiation) of tissues of two different individuals so that one (the scion) can survive on the other (the stock).

ground tissue: tissue (usually parenchyma) surrounding vascular tissue, and within the epidermis.

growth ring: (in secondary xylem) a distinct growth increment, caused by differential rates of growth during a growing season.

guard cell: one of a pair of cells of a stoma, surrounding pore.

guttation: secretion (usually passive) of water droplets, often at hydathodes.

gynoecium: (in flower) carpels; including ovary, style(s) and stigma(s).

hair: epidermal appendage, or trichome.

haustorium: (in parasitic plants) modified root through which nutrients pass from host to parasite.

(in endosperm/embryo) specialised absorptive cells of embryo sac, endosperm or suspensor, that may penetrate into adjacent tissues.

hemicellulose: carbohydrate constituent of plant cell walls; less common than cellulose.

heterocellular ray: one composed of cells of different shapes and sizes.

hilum: scar on seed indicating point of attachment of funicle to ovary wall.

histogenesis: tissue differentiation.

homocellular ray: one composed of cells which are the same shape and size.

homologous: (in the morphological sense) having a similar ontogenetic sequence.

hydathode: region of secretion of water droplets (usually on margins of leaves).

hydrophyte: water plant, sometimes displaying hydromorphic features (e.g. aerenchyma).

hypocotyl: seedling axis bearing cotyledons and shoot apex.

hypodermis: (often applied to leaves) distinct cell layer(s) immediately within epidermis, other than "normal" photosynthetic mesophyll.

hypogeal germination: type where cotyledons remain enclosed in seed coat after radicle has emerged.

hypogynous flower: one with superior ovary; i.e. above the level of insertion of the stamens and perianth parts.

hypostase: proliferation of nucellus at chalazal end of embryo sac, often bearing antipodals (Fig. 5.13b).

hypostomatic leaves: with stomata found on abaxial surface only.

idioblast: cell that differs from the cells of the surrounding tissue.

included phloem: areas of phloem embedded in secondary xylem.

integument: either one or two (inner and outer) structures ensheathing the embryo sac, around the nucellus (Figs 5.12, 5.13).

intercalary growth, or meristem: cell divisions in region apart from apical meristem or other well-defined meristems.

internode: region between two nodes on stem.

intine: inner (usually softer) part of pollen grain wall.

isobilateral leaf: one with both surfaces similar, or with palisade tissue on both sides (abaxially and adaxially).

Kranz anatomy: (in some plants with C4 photosynthesis) distinctive leaf anatomy with mesophyll cells radiating from the vascular bundle sheaths, which are usually a single layer of parenchymatous cells containing starch and chloroplasts.

lateral roots: branches of the tap root; they have an endogenous origin.

latex: fluid contained within laticifers, consisting of a suspension of fine particles; used to form rubber.

laticifer: latex-secreting cell.

lenticel: region of loose cells in the periderm (bark).

lignin: substance often deposited in cell walls of strengthening tissues (e.g. fibres), giving rigidity.

limb: part of monocotyledon seedling (Fig. 6.7).

macrosclereid: elongated sclereid often found in seed coat.

megagametophyte: mature embryo sac, most commonly consisting of eight nuclei (two synergids, an egg cell, two polar nuclei and three antipodals).

megaspore: female haploid cell resulting from meiosis; usually one of two or four, of which only one is functional.

megasporogenesis: process of megaspore formation from a megaspore mother cell.

meiosis: two successive divisions of a diploid nucleus to form a haploid gamete.

meristem: region of cell division and tissue differentiation (e.g. apical meristem, intercalary meristem, lateral meristem, vascular cambium, primary and secondary thickening meristems).

meristemoid: isolated meristematic cell or group of cells (e.g. guard cell mother cell).

mesocarp: middle layer of pericarp (fruit wall).

mesogene cell: stomatal subsidiary cell derived from meristemoid.

mesomorphic: displaying no xeromorphic or hydromorphic characteristics.

mesophyll: ground tissue of leaf; mainly consisting of parenchyma or chlorenchyma; often differentiated into palisade and spongy mesophyll.

mesophyte: plant growing in conditions of fairly continuous moisture.

metaxylem: primary xylem formed after protoxylem.

microfibril: thread-like component of cell wall, usually of cellulose.

micropyle: opening at one end of the ovule, usually surrounded by the integuments (Fig. 5.13).

microsporangium: pollen sac, contained within anther.

microspore: male spore that will give rise to male gametophyte (pollen grain). Undergoes unequal division to form vegetative nucleus and generative nucleus.

microsporogenesis: process of microspore formation.

middle lamella: layer between the walls of neighbouring cells.

mitochondrion pl. **mitochondria**: cytoplasmic organelle.

mitosis: cell division to form two cells of equivalent chromosome composition to parent cell; involving four main stages: prophase, metaphase, anaphase and telophase.

mucilage (slime): compound that swells in water, occurring in some plant cell walls, such as gelatinous or mucilaginous cells.

multiseriate: consisting of more than one layer or row of cells.

nectary: (floral or extrafloral) localised cell or cells that secrete a sugary liquid (nectar).

nexine: inner, non-sculptured part of exine.

node: part of stem where leaves are attached.

non-articulated laticifer: one composed of a single multinucleate coenocytic cell.

nucellus: ovule cell layer(s) immediately surrounding embryo sac (megagametophyte).

obturator: proliferation of (usually) ovary tissue near micropyle, with secretory function, to guide growing pollen tubes into micropyle.

ontogeny: development; differentiation and growth.

organogenesis: development of organs.

osmophore: scent-producing gland in flower.

osteosclereid: bone-shaped sclereid (Fig. 4.9).

papilla (pl. **papillae**): epidermal appendage; small unicellular trichome.

paracytic: one of the classification types of mature stomata; with, at either side of the guard cells, one or more subsidiary cells that may or may not meet at the poles.

paratracheal parenchyma: (in secondary xylem) associated with the vessels.

parenchyma: tissue composed of least specialised cell type, thin-walled cells with living contents.

passage cell: an endodermal cell that remains thinner walled than neighbouring endodermal cells.

pearl glands (pearl bodies): secretory, often globular, leaf emergences or trichomes that provide food (carbohydrates, lipids and proteins) for ants, which carry them off.

perforated ray cell: (in secondary xylem) ray cell linking two vessel elements, and itself resembling and functioning as a vessel element.

perforation plate: opening in end wall of vessel element; may be simple (a single opening) or scalariform (with bars: Fig. 1.8), or more rarely, reticulate (mesh-like) or foraminate (with pores).

perianth: outer sterile part of flower, consisting of whorls of sepals (calyx) and petals (corolla), or sometimes undifferentiated tepals.

pericarp: fruit wall.

periclinal: parallel with the surface.

pericycle: in roots, a usually distinct layer of thin-walled cells within the endodermis; in stems pericyclic region (surrounding more central vascular region) often less clearly defined or absent.

periderm: cork tissue.

perigene cell: stomatal subsidiary cell derived from cell adjacent to meristemoid, not from the meristemoid itself.

periplasmodium: coalescent mass in anther locule, formed from protoplasts of tapetal cells.

perisperm: food-storage tissue in the seed, derived from part of the nucellus.

phellem: cork cambium, or cork meristem.

phelloderm: internal derivatives of phellem.

phellogen: external derivatives of phellem.

phloem: tissue that transports food in the form of assimilates. May be primary (produced by the apical meristem) or secondary (produced by the vascular cambium).

phyllotaxis: the pattern of arrangement of organs on an axis, e.g. leaves on a stem, flowers on an inflorescence.

pith: central parenchymatous region of stems, often breaking down to form a cavity.

pits: thin areas of the primary and secondary cell wall, often corresponding with pits in adjacent cells (pit-pairs).

placenta: region of attachment of ovules on ovary wall.

placentation: arrangement of placentae and locules in ovary (e.g. axile, basal, free central, parietal).

plasma membrane (plasmalemma): cell membrane (within cell wall) that encloses protoplast.

plasmodesmata: protoplasmic strands passing through primary pit fields between adjacent cells, and connecting their protoplasts.

plastid: cell organelle contained within cytoplasm, often with specialised function (e.g. chloroplast, amyloplast).

polar nucleus: one of a pair of nuclei of the mature embryo sac (megagametophyte), often in a central position. The two haploid polar nuclei fuse to form a diploid fusion nucleus, which further fuses with a male haploid nucleus to form the triploid endosperm.

pollen grain: male gametophyte; bicellular (or rarely tricellular) at maturity.

pollen tube: tube emerging from germinating pollen grain on stigma, which will transport male nuclei to megagametophyte.

polyarch root: one with several (more than four) protoxylem poles.

primordium: young, newly differentiating organ.

procambium: primary tissue near (shoot or root) apex, that gives rises to primary vascular tissue.

proembryo: young globular embryo, prior to differentiation of cotyledons and hypocotyl.

promeristem: (in root apices) region of greatest mitotic activity.

protoplast: living part of cell, surrounded by a plasma membrane.

protoxylem: first-formed primary xylem.

proximal: situated closer to the centre of a body or its point of attachment; opposite of distal.

quiescent centre: region of cells at root apex, with little or no cell division activity (Fig. 3.1).

radicle: first-formed root of seedling.

raphe: stalk attaching seed to ovary (directly derived from the funicle).

raphide: fine, needle-like crystal, one of a group of several raphides formed within a single cell.

ray: (in secondary xylem) tissue of radially oriented cells, usually parenchymatous.

root cap: protective covering of cells over root apex.

root hair: water-absorbing hair on root epidermis.

sarcotesta: fleshy part of seed coat.

scale (peltate hair): a modified trichome, consisting of a fused disc of cells attached to the epidermis by a stalk.

sclereid: thick-walled sclerenchymatous cell, usually lacking contents at maturity, and either isodiametric (brachysclereid or stone cell), or variously shaped (e.g. astrosclereid, osteosclereid, macrosclereid).

sclerenchyma: strengthening tissue, consisting of cells with thickened, lignified walls, usually lacking contents at maturity. Includes fibres and sclereids.

scutellum: specialised structure in grass embryos, often interpreted as a modified cotyledon (Fig. 6.6).

sexine: outer, sculptured part of exine.

sieve tube element: phloem conducting cell. Cell wall with sieve areas, which are mainly localised on end walls adjoining adjacent sieve elements, in the form of sieve plates. Groups of sieve tube elements are linked axially to form sieve tubes.

stigma: (on style) secretory tissue that is receptive to pollen grains.

stoma (pl. **stomata**): pore in epidermis for gaseous exchange, usually on aerial parts of plant (rarely on roots); consisting of two guard cells surrounding central pore.

stone cell: isodiametric sclereid.

storied structure: (in secondary xylem, usually referring to vessels, rays or axial parenchyma) stratified, or occurring in rows.

styloid: elongated prismatic crystal.

suberin: fatty compound sometimes deposited in cell walls, e.g. in cork cells.

subsidiary cells: epidermal cells adjacent to stomata or other subsidiary cells, that differ from surrounding epidermal cells.

sulcus: aperture in pollen grain wall on its distal face (i.e. the face that is directed outwards in the tetrad).

suspensor: row of cells attaching globular proembryo to wall of embryo sac (i.e. embryo stalk).

symplast: connected living protoplasts of adjacent cells.

syncarpous gynoecium: one where carpels are more or less fused, although stigmas and styles may remain separate.

synergid: one of a pair of cells at micropylar end of mature embryo sac (megagametophyte).

tannins: a broad group of phenol derivatives. Tannin derivatives appear as a brown, yellow or red amorphous substance in cells of sectioned material.

tap root: main central root, formed directly from seedling radicle.

tapetum: innermost layer of cells of the anther wall, immediately surrounding the anther locule (Fig. 5.5).

tegmen: the inner layer of the seed coat, formed from the inner integument.

testa: seed coat; or in bitegmic seeds, the outer layer of the seed coat, formed from the outer integument.

tetrad: spores (either megaspores or microspores) united in group of four.

tetrarch root: one with four protoxylem poles.

tracheid: xylem water-conducting cell, usually with bordered pits, but lacking perforation plates.

transmitting tissue: (or stigmatoid tissue) secretory tissue through which pollen tubes grow, linking stigma with centre of ovary.

triarch root: one with three protoxylem poles.

trichoblast: root epidermal cell with dense cytoplasm, that will give rise to a root hair.

trichome: epidermal outgrowth (hair).

tunica-corpus: theory of shoot apex organisation; the tunica being the outermost (usually one to six) cell layers, with mainly anticlinal cell divisions, and the corpus the cell layers within them, with randomly oriented divisions.

tylose: (in secondary xylem) outgrowth of the wall of an axial parenchyma cell into a vessel element through a pit; eventually often completely blocking vessel.

unifacial leaf: one with both surfaces similar, sometimes derived from a single (usually abaxial) surface.

uninterrupted meristem: region of diffuse cell divisions that is continuous with the apical meristem; producing extension growth of the axis.

uniseriate: consisting of a single layer or row of cells.

vacuole: cavity.

vascular bundle: strand of vascular (conducting) tissue.

vascular cambium: meristem that produces secondary vascular tissue in dicotyledons; initially normally situated between primary xylem and phloem in vascular bundles.

vascular tissue: conducting tissue (phloem and xylem).

vegetative cell: one of two cells of male gametophyte.

velamen: multiseriate (several layered) epidermis on the aerial roots of some tropical epiphytes, such as orchids and aroids.

venation: arrangement of vascular bundles in leaf (e.g. parallel venation, or reticulate venation).

vessel element: water-conducting cell of xylem, with bordered pits on side walls and perforation plates on end walls adjacent to adjoining vessel elements; groups of axially linked vessel elements form a vessel.

vestured pitting: (in secondary xylem of certain dicotyledons) bordered pits surrounded by numerous warty protuberances.

wax: fatty substance often deposited on the surface of the cuticle.

wood: secondary xylem.

xeromorphic: showing characteristics that are often associated with dry environments.

xerophyte: plant that grows in a dry (xeric) environment.

xylem: water-transporting tissue, consisting of several different cell types. Primary xylem (protoxylem and metaxylem) is formed by the apical meristem. Secondary xylem is formed by the vascular cambium.

zygote: cell formed by fusion of female egg cell and male sperm cell; will eventually divide to form the proembryo.

Further general reading

ARBER, A. (1925). *Monocotyledons: a morphological study.* Cambridge University Press, Cambridge.

BELL, A. D. (1991). *Plant form.* Oxford University Press, Oxford.

CORNER, E. J. H. (1976). *The seeds of Dicotyledons.* Cambridge University Press, Cambridge.

ESAU, K. (1965). *Plant anatomy.* 2nd ed. John Wiley and Sons, New York.

FAHN, A. (1979). *Secretory tissues in plants.* Academic Press, London.

FAHN, A. (1990). *Plant anatomy.* 4th ed. Pergamon Press, Oxford.

JANE, F. W. (1970). *The structure of wood.* 2nd ed. Adam and Charles Black, London.

JOHRI, B. M. (ED.) (1984). *Embryology of Angiosperms.* Springer-Verlag, Berlin.

MAHESHWARI, P. (1950). *An introduction to the embryology of the Angiosperms.* McGraw-Hill Book Company, New York.

METCALFE, C. R. and CHALK, L. (1979 and 1983). *Anatomy of the Dicotyledons.* 2nd ed., Vols I and II. Clarendon Press, Oxford.

STEEVES, T. A. and SUSSEX, I. M. (1989). *Patterns in plant development.* Cambridge University Press, Cambridge.

WEBERLING, F. (1981). *Morphology of flowers and inflorescences.* Translated by R. J. Pankhurst, 1989. Cambridge University Press, Cambridge.

References

ARBER, A. (1925). *Monocotyledons: a morphological study*. Cambridge University Press, Cambridge.

BAAS, P. AND GREGORY, M. (1985). A survey of oil cells in the dicotyledons with comments on their replacement by and joint occurrence with mucilage cells. *Israeli Journal of Botany* **34**: 167–186.

BARLOW, P. W. (1975). The root cap. In: *The development and form of roots*. J. G. Torrey and D. F.Clarkson (Eds). Academic Press, New York and London.

BARTHLOTT, W. (1981). Epidermal and seed surface characters of plants: systematic applicability and some evolutionary aspects. *Nordic Journal of Botany* **1**: 345–355.

BARTHLOTT, W. AND EHLER, N. (1977). Raster-Elektromikroskopie der Epidermis-Oberflächen von Spermatophyten. *Tropische und Subtropische Pflanzenwelt* **19**: 1–110.

BARTHLOTT, W. AND WOLLENWEBER, E. (1981). Zur Feinstruktur, Chemie und taxonomischen Signifikanz epicuticularer Wachse und ahnlicher Secrete. *Tropische und Subtropische Pflanzenwelt* **32**: 1–67.

BEHNKE, H.-D. (1972). Sieve tube plastids in relation to angiosperm systematics – an attempt toward classification by ultrastructural analysis. *Botanical Review* **38**: 155–197.

BEHNKE, H.-D. (1975). The basis of angiosperm phylogeny: ultrastructure. *Annals of the Missouri Botanical Garden* **62**: 647–663.

BEHNKE, H.-D. (1981). Sieve element characters. *Nordic Journal of Botany* **1**: 381–400.

BÖCHER, T. W. AND LYSHEDE, O. B. (1968). Anatomical studies on xerophytic apophyllous plants. I. *Montea aphylla*, *Bulnesia retama* and *Bredemeyera colletioides*. *Biologiscke Skrifter* **16**: 1–44.

BÖCHER, T. W. AND LYSHEDE, O. B. (1972). Anatomical studies on xerophytic apophyllous plants. II. Additional species from South American shrub steppes. *Biologiscke Skrifter* **18**: 1–137.

BOESEWINKEL, F. D. AND BOUMAN, F. (1984). The seed: structure. In: *Embryology of Angiosperms*. B. M. Johri (Ed.), Springer-Verlag, Berlin.

BRANDHAM, P. E. AND CUTLER, D. F. (1978). Influence of chromosome variation on the organisation of the leaf epidermis in a hybrid *Aloë* (Liliaceae). *Botanical Journal of the Linnean Society* **77**: 1–16.

BURGESS, J. (1985). *An introduction to plant cell development*. Cambridge University Press, Cambridge.

CLOWES, F. A. L. (1961). *Apical meristems*. Blackwell, Oxford.

CORNER, E. J. H. (1976). *The seeds of Dicotyledons*. Cambridge University Press, Cambridge.

CUTLER, D. F. AND BRANDHAM, P. E. (1977). Experimental evidence for the genetic control of leaf surface characters in hybrid Aloineae (Liliaceae). *Kew Bulletin* **32**: 23–42.

CUTLER, D. F., RUDALL, P. J., GASSON, P. E. AND GALE, R. M. O. (1987). *Root identification manual of trees and shrubs*. Chapman and Hall, London.

DAUMANN, E. (1970). Das Blutennektarium der Monocotyledon unter besonderer Berucksichtigung seiner systematischen und phylogenetischen Bedeutung. *Feddes Repertorium* **80**: 463–590.

DAVIS, G. L. (1966). *Systematic anatomy of the Angiosperms*. John Wiley and Sons, New York.

DEMASON, D. A. (1983). The primary thickening meristem: definition and function in monocotyledons. *American Journal of Botany* **70**: 955–962.

DIGGLE, P. K. AND DEMASON, D. A. (1983). The relationship between the primary thickening meristem and the secondary thickening meristem in *Yucca whipplei* Torr. I. Histology of the mature vegetative stem. *American Journal of Botany* **70**: 1195–1204.

EAMES, A. J. AND MACDANIELS, L. H. (1925). *An introduction to plant anatomy*. McGraw-Hill Book Company, New York.

ENDRESS, P. K., JENNY, M. AND FALLEN, M. E. (1983). Convergent elaboration of apocarpous gynoecia in higher advanced dicotyledons (Sapindales, Malvales, Gentianales). *Nordic Journal of Botany* **3**: 293–300.

ERDTMAN, G. (1966). *Pollen morphology and plant taxonomy*. Hafner Publishing Company, New York and London.

ESAU, K. (1965). *Plant anatomy*. 2nd ed. John Wiley and Sons, New York.

FAHN, A. (1979). *Secretory tissues in plants*. Academic Press, London.

FELDMAN, L. J. (1984). The development and dynamics of the root apical meristem. *American Journal of Botany* **71**: 1308–1314.

FISHER, J. B. AND FRENCH, J. C. (1976). The occurrence of intercalary and uninterrupted meristems in the internodes of tropical monocotyledons. *American Journal of Botany* **63**: 510–525.

FISHER, J. B. AND FRENCH, J. C. (1978). Internodal meristems of monocotyledons: further studies and a general taxonomic summary. *Annals of Botany* **42**: 41–50.

FOSTER, A. S. (1947). Structure and ontogeny of the terminal sclereids in the leaf of *Mouriria huberi* Cogn. *American Journal of Botany* **34**: 501–514.

FOSTER, A. S. (1956). Plant idioblasts; remarkable examples of cell specialisation. *Protoplasma* **46**: 184–193.

FRANCESCHI, V. R. AND HORNER, H. T. (1980). Calcium oxalate crystals in plants. *Botanical Review* **46**: 361–427.

FRENCH, J. C. (1986). Patterns of stamen vasculature in the Araceae. *American Journal of Botany* **73**: 434–449.

GALE, R. M. O. AND OWENS, S. J. (1983). Cell distribution and surface morphology in petals, androecia and styles of Commelinaceae. *Botanical Journal of the Linnean Society* **87**: 247–262.

GIFFORD, E. M. AND CORSON, G. E. (1971). The shoot apex in seed plants. *Botanical Review* **37**: 147–229.

GREGORY, M. AND BAAS, P. (1989). A survey of mucilage cells in vegetative organs of the dicotyledons. *Israeli Journal of Botany* **38**: 125–174.

HESLOP-HARRISON, J. (1976). The adaptive significance of the exine. In: *The evolutionary significance of the exine*. I. K. Ferguson and J. Muller (Eds). Academic Press, London.

HESLOP-HARRISON, J. AND HESLOP-HARRISON, Y. (1982). The specialised cuticles of the receptive surfaces of angiosperm stigmas. In: *The plant cuticle*. D. F. Cutler, K. L. Alvin and C. E. Price (Eds). Academic Press, London.

HESLOP-HARRISON, J. AND SHIVANNA, K. R. (1977). The receptive surface of the angiosperm stigma. *Annals of Botany* **41**: 1233–1258.

HICKEY, L. J. (1973). A classification of the architecture of dicotyledonous leaves. *American Journal of Botany* **60**: 17–33.

HICKEY, L. J. (1979). A revised classification of the architecture of dicotyledonous leaves. In: *Anatomy of the Dicotyledons.* C. R. Metcalfe and L. Chalk (Eds). 2nd ed., Vol I. Clarendon Press, Oxford.

HOWARD, R. A. (1974). The stem–node–leaf continuum of the Dicotyledonae. *Journal of the Arnold Arboretum* **55**: 125–181.

HOWARD, R. A. (1979). The stem–node–leaf continuum of the Dicotyledonae. In: *Anatomy of the Dicotyledons.* C. R. Metcalfe and L. Chalk (Eds). 2nd ed., Vol I. Clarendon Press, Oxford.

HUYNH, K. L. (1976). Arrangement of some monosulcate, disulcate, trisulcate, dicolpate and tricolpate pollen types in the tetrads, and some aspects of evolution in the Angiosperms. In: *The evolutionary significance of the exine.* I. K. Ferguson and J. Muller (Eds) Academic Press, London.

JERNSTEDT, J. A. (1984). Root contraction in hyacinth. I. Effects of IAA on differential cell expansion. *American Journal of Botany* **71**: 1080–1089.

JUNIPER, B. E. AND JEFFREE, C. E. (1983). *Plant surfaces.* Edward Arnold, London.

KAPLAN, D. R. (1970). Comparative foliar histogenesis in *Acorus calamus* and its bearing on the phyllode theory of monocotyledonous leaves. *American Journal of Botany* **57**: 331–361.

KAPLAN, D. R. (1973a). The Monocotyledons: their evolution and comparative biology. VII. The problem of leaf morphology and evolution in the Monocotyledons. *Quarterly Review of Biology* **48**: 437–451.

KAPLAN, D. R. (1973b). Comparative developmental analysis of the heteroblastic leaf series of axillary shoots of *Acorus calamus* L. (Araceae). *Cellule* **69**: 251–290.

KAPLAN, D. R. (1975). Comparative developmental evaluation of the morphology of unifacial leaves in the monocotyledons. *Botanische Jahrbücher* **95**: 1–105.

KAPLAN, D. R. (1984). Alternative modes of organogenesis in higher plants. In: *Contemporary problems in plant anatomy.* R. A. White and W. C. Dickinson (Eds). Academic Press, New York and London.

KAY, Q. O. N., DAOUD, H. S. AND STIRTON, C. H. (1981). Pigment distribution, light reflection and cell structure in petals. *Botanical Journal of the Linnean Society* **83**: 57–84.

KNOX, R. B. (1984). The pollen grain. In: *Embryology of Angiosperms.* B. M. Johri (Ed.). Springer-Verlag, Berlin.

KUIJT, J. (1969). *The biology of parasitic flowering plants.* University of California Press, Berkeley and Los Angeles.

MAHESHWARI, P. (1950). *An introduction to the embryology of the Angiosperms.* McGraw-Hill Book Company, New York.

MAHLBERG, P. (1975). Evolution of the laticifer in *Euphorbia* as interpreted from starch grain morphology. *American Journal of Botany* **62**: 577–583.

MCCULLY, M. E. (1975). The development of lateral roots. In: *The development and function of roots.* J. G. Torrey and D. T. Clarkson (Eds). Academic Press, London.

METCALFE, C. R. AND CHALK, L. (1979 and 1983). *Anatomy of the Dicotyledons.* 2nd ed., Vols I and II. Clarendon Press, Oxford.

MILES, A. (1978). *Photomicrographs of world woods.* HMSO, London.

NATESH, S. AND RAU, M. A. (1984). The embryo. In: *Embryology of Angiosperms.* B. M. Johri (Ed.). Springer-Verlag, Berlin.

OBATON, M. (1960). Les lianes ligneuses structure anormale des forts denses d'Afrique occidentale. *Annales des Sciences Naturelles, Botanique* ser.12, **1**: 1–120.

O'DOWD, D. J. (1982). Pearl bodies as ant food: an ecological role for some leaf emergences of tropical plants. *Biotropica* **14**: 40–49.

PURI, V. (1951). The role of floral anatomy in the solution of morphological problems. *Botanical Review* **17**: 471–553.

RASMUSSEN, H. (1983). Stomatal development in families of Liliales. *Botanische Jahrbücher* **104**: 261–287.

RUDALL, P. (1991). Lateral meristems and stem thickening growth in monocotyledons. *Botanical Review* **57**: 150–163.

RUZIN, S. E. (1979). Root contraction in *Freesia* (Iridaceae). *American Journal of Botany* **66**: 522–531.

SATTLER, R. (1973). *Organogenesis of flowers. A photographic atlas.* University of Toronto Press, Toronto.

SCHENCK, H. (1893). *Beitrage zur Anatomie der Lianen. Schimper's Botanischer Mittelheilungen aus den Tropen,* Heft 5.

SCHMID, R. (1972). Floral bundle fusion and vascular conservatism. *Taxon* **21**: 429–446.

SCHMID, R. (1988). Reproductive versus extra-reproductive nectaries – historical perspective and terminological recommendations. *Botanical Review* **54**: 179–232.

SCHMIDT, A. (1924). Histologische Studien an phanerogamen Vegetationspunkten. *Botanisches Archiv* **8**: 345–404.

SCHWEINGRUBER, F. H. (1990). *Anatomy of European woods*. Haupt, Bern and Stuttgart.

SHIGO, A. L. (1985). How tree branches are attached to trunks. *Canadian Journal of Botany* **63**: 1391–1401.

STEEVES, T. A. AND SUSSEX, I. M. (1989). *Patterns in plant development*. Cambridge University Press, Cambridge.

STERLING, C. (1972). Mechanism of root contraction in *Gladiolus*. *Annals of Botany* **36**: 589–598.

STEVENSON, D. W. (1980). Radial growth in *Beaucarnea recurvata*. *American Journal of Botany* **67**: 476–489.

STEVENSON, D. W. AND FISHER, J. B. (1980). The developmental relationship between primary and secondary thickening growth in *Cordyline* (Agavaceae). *Botanical Gazette* **141**: 264–268.

STEVENSON, D. W. AND OWENS, S. J. (1978). Some aspects of the reproductive morphology of *Gibasis venustula* (Kunth) D. R. Hunt (Commelinaceae). *Botanical Journal of the Linnean Society* **77**: 157–175.

TILLICH, H. J. (1977). Vergleichend-morphologische Untersuchungen zur Identität der Gramineen-Primärwurzel. (Comparative morphological investigations on the identity of the primary root of the Gramineae). *Flora* **166**: 415–421.

TILTON, V. R. AND HORNER, H. T. (1980). Stigma, style and obturator of *Ornithogalum caudatum* (Liliaceae) and their function in the reproductive process. *American Journal of Botany* **67**: 1113–1131.

TOMLINSON, P. B. (1974). Development of the stomatal complex as a taxonomic character in Monocotyledons. *Taxon* **23**: 109–128.

TOMLINSON, P. B. AND ZIMMERMAN, M. H. (1969). Vascular anatomy of monocotyledons with secondary growth – an introduction. *Journal of the Arnold Arboretum* **50**: 159–179.

TUCKER, S. C. (1972). The role of ontogenetic evidence in floral morphology. In: *Advances in plant morphology*. Y. F. Murty *et al.* (Eds). Sarita Prakashan, Meerut, India.

UHL, N W. AND MOORE, H. E. (1977). Centrifugal stamen initiation in phytelephantoid palms. *American Journal of Botany* **64**: 1152–1161.

VIJAYARAGHAVAN, M. R. AND PRABHAKAR, K. (1984). The endosperm. In: *Embryology of Angiosperms*. B. M. Johri (Ed.). Springer-Verlag, Berlin.

VOGEL, S. (1961). Osmophoren. Über einen neuartigen Typus pflanzlichen Drüsengewebes. *Bericht der Deutschen Botanischen Gesellschaft* **73**.

VOGEL, S. (1974). Olbumen und olsammelnde Bienen. *Tropische und Subtropische Pflanzenwelt* **7**: 285–577.

WILKINSON, H.P. (1979). The plant surface. In: *Anatomy of the Dicotyledons*. C.R. Metcalfe and L. Chalk (Eds). 2nd ed., Vol I. Clarendon Press, Oxford.

WILLEMSE, M.T.M. AND VAN WENT, J.L. (1984). The female gametophyte. In: *Embryology of Angiosperms*. B.M. Johri (Ed.). Springer-Verlag, Berlin.

WILSON, K. AND ANDERSON, G.J.H. (1979). Further observations on root contraction. *Annals of Botany* **43**: 665–675.

ZIMMERMAN, M.H. AND TOMLINSON, P.B. (1965). Anatomy of the palm *Rhapis excelsa*. I. Mature vegetative axis. *Journal of the Arnold Arboretum* **46**: 160–180.

ZIMMERMAN, M.H. AND TOMLINSON, P.B. (1966). Analysis of complex vascular systems in plants: optical shuttle method. *Science* **152**: 122–142.

ZIMMERMAN, M.H. AND TOMLINSON, P.B. (1972). The vascular system of monocotyledonous stems. *Botanical Gazette* **133**: 141–155.

Index

Numbers in bold indicate illustrations